Electrical characterization of transition metals
in silicon: a study on titanium, cobalt, and
nickel and their interaction with hydrogen

AF303155

Electrical characterization of transition metals in silicon: a study on titanium, cobalt, and nickel and their interaction with hydrogen

Der Fakultät Mathematik und Naturwissenschaften der
Technischen Universität Dresden

zur Erlangung des akademischen Grades eines

Doctor rerum naturalium

eingereichte Dissertation

von

Herrn Dipl.-Phys. Leopold Scheffler

geboren am 27. 06. 1985 in Berlin

Bibliografische Information der Deutschen Nationalbibliothek

Die Deutsche Nationalbibliothek verzeichnet diese Publikation in der Deutschen Nationalbibliografie; detaillierte bibliografische Daten sind im Internet über http://dnb.d-nb.de abrufbar.

    1. Aufl. - Göttingen: Cuvillier, 2015

    Zugl.: (TU) Dresden, Univ., Diss., 2015

Datum des Einreichens: 18.12.2014

Datum der Verteidigung: 01.04.2015

1. Gutachter: Prof. Dr. Jörg Weber

2. Gutachter: Dir. of Res. Dr. Abdelmadjid Mesli

© CUVILLIER VERLAG, Göttingen 2015
    Nonnenstieg 8, 37075 Göttingen
    Telefon: 0551-54724-0
    Telefax: 0551-54724-21
    www.cuvillier.de

Alle Rechte vorbehalten. Ohne ausdrückliche Genehmigung des Verlages ist es nicht gestattet, das Buch oder Teile daraus auf fotomechanischem Weg (Fotokopie, Mikrokopie) zu vervielfältigen.

1. Auflage, 2015

Gedruckt auf umweltfreundlichem, säurefreiem Papier aus nachhaltiger Forstwirtschaft

    ISBN 978-3-95404-988-2
    eISBN 978-3-7369-4988-1

Teile der vorliegenden Arbeit wurden bereits veröffentlicht:

- L. Scheffler, Vl. Kolkovsky und J. Weber, *Identification of titanium-hydrogen complexes with up to four hydrogen atoms in silicon*, J. Appl. Phys. **117**, 085707 (2015)

- L. Scheffler, Vl. Kolkovsky und J. Weber, *Isolated Ti in Si: Deep level transient spectroscopy, minority carrier transient spectroscopy, and high-resolution Laplace deep level transient spectroscopy studies*, J. Appl. Phys. **117**, 045713 (2015)

- L. Scheffler, Vl. Kolkovsky und J. Weber, *Electrical Levels in Nickel doped Silicon*, J. Appl. Phys. **116**, 173704 (2014)

- L. Scheffler, Vl. Kolkovsky und J. Weber, *Has substitutional nickel two acceptor- and one donor-level in Si?*, AIP Conf. Proc. **1583**, 85 – 89 (2014)

- Vl. Kolkovsky, L. Scheffler und J. Weber, *Transition metals (Ti and Co) in silicon and their complexes with hydrogen: A Laplace DLTS study*, Physica B **439**, 24 – 28 (2014)

- J. Weber, L. Scheffler, Vl. Kolkovsky und N. Yarykin, *New Results on the Electrical Activity of 3d-Transition Metal Impurities in Silicon*, Solid State Phenom. **205-206**, 245 (2013)

- L. Scheffler, Vl. Kolkovsky und J. Weber, *Isolated substitutional cobalt and Co-related complexes in silicon*, J. Appl. Phys. **113**, 183714 (2013)

- L. Scheffler, Vl. Kolkovsky und J. Weber, *A re-examination of cobalt-related defects in n- and p-type silicon*, Phys. Status Solidi A **209**, 1913 – 1916 (2012)

- Vl. Kolkovsky, L. Scheffler und J. Weber, *A re-examination of the interstitial Ti levels in Si*, Phys. Status Solidi C **9**, 1996 – 1999 (2012)

Abstract

The understanding of the electrical properties of defects introducing deep levels in silicon is of prime technological importance in modern microelectronics. In this thesis, a comprehensive study of the transition metals titanium, cobalt, and nickel in silicon, and of their interaction with hydrogen is presented. The formed defects are detected and characterized by deep level transient spectroscopy (DLTS), Laplace DLTS, and minority carrier transient spectroscopy.

A natural starting point for a study of metal-hydrogen reactions in silicon is the analysis of the effect of hydrogen on metal-free silicon. Complexes of hydrogen with carbon, which create deep levels in the band gap of silicon, are observed.

Titanium introduces three levels into the band gap. The charge states determined in this thesis are in contradiction to the literature, questioning the assignment of these levels. Upon hydrogenation, TiH complexes with one, two, and three hydrogen atoms are identified. A proposition by theory that two different configurations of TiH with one hydrogen atom exist, can be supported.

Cobalt is shown to have only one level in the band gap of silicon, whereas a second level previously attributed to cobalt is assigned to the cobalt-boron pair. Two CoH complexes are determined.

Nickel has three levels in the band gap. Upon hydrogenation, complexes with up to three hydrogen atoms are identified. One of the defects can be observed in both n- and p-type silicon.

For all three metals investigated, passive hydrogen complexes exist. They are created by further hydrogenation after the appearance of the above mentioned electrically active complexes. The thesis concludes with a comparison of the obtained results with those of neighboring elements to look for similarities and patterns.

Kurzfassung

Das Verständnis der elektrischen Eigenschaften von Defekten, welche tiefe Niveaus in der Bandlücke von Silizium erzeugen, ist von außerordentlichem Interesse für die moderne Mikroelektronik. In der vorliegenden Dissertation wird eine umfassende Untersuchung der Übergangsmetalle Titan, Kobalt und Nickel in Silizium und ihrer Wechselwirkung mit Wasserstoff vorgestellt. Die entstandenen Defekte werden mit Hilfe von Kapazitätstransientenspektroskopie (DLTS - deep level transient spectroscopy), Laplace DLTS und Minoritätsladungsträgertransientenspektroskopie (MCTS - minority carrier transient spectroscopy) beobachtet und charakterisiert.

Für eine fehlerfreie Analyse der Metall-Wasserstoff-Reaktionen ist es sinnvoll, zuerst den Einfluss des Wasserstoffs auf metallfreies Silizium zu prüfen. Dabei wird die Bildung von Kohlenstoff-Wasserstoff-Komplexen, welche Niveaus in der Bandlücke von Silizium erzeugen, beobachtet.

Titan besitzt drei Niveaus in der Bandlücke von Silizium. Die in dieser Arbeit bestimmten Ladungszustände stehen im Widerspruch zu den Literaturangaben, daher wird die Zuordnung dieser Niveaus in Frage gestellt. Die Reaktion von Titan mit Wasserstoff führt zu elektrisch aktiven Komplexen mit bis zu drei Wasserstoffatomen. Die Ergebnisse unterstützen einen Vorschlag aus der Theorie, nach dem der Komplex mit einem Wasserstoff zwei verschiedene Konfigurationen besitzen soll.

Kobalt erzeugt ein Niveau in der Bandlücke. Ein weiteres Niveau, welches früher ebenfalls dem Kobalt zugewiesen wurde, kann dem Kobalt-Bor-Paar zugeordnet werden. Nach der Reaktion mit Wasserstoff können zwei CoH-Komplexe nachgewiesen werden.

Nickel besitzt drei Niveaus in der Bandlücke und erzeugt elektrisch aktive NiH-Komplexe mit bis zu drei Wasserstoffatomen. Einer dieser Defekte kann sowohl im n- als auch im p-Typ Silizium beobachtet werden.

Alle drei untersuchten Metalle besitzen elektrisch passive Komplexe, welche nach der weiteren Reaktion von Wasserstoff mit den aktiven Komplexen entstehen. Die Arbeit endet mit einem Vergleich der Ergebnisse mit denen benachbarter Elemente, um mögliche Ähnlichkeiten oder Muster zu erkennen.

Contents

Chapter 1

Introduction

High-purity silicon is an important resource in our modern world. Products based on silicon permeate our daily life, from integrated circuits in desktop computers, laptops, and mobile devices, to sensors in our cars, homes, and industrial production facilities, to photovoltaic systems on our roofs. Since the beginning of mass production of integrated circuits in the 1970s, research on the properties of defects in silicon has been a key to the evolution of present-day applications.

Electrically active defects which create deep levels in the electronic band gap of silicon fundamentally influence device performance. These defects can increase the recombination activity of electron-hole pairs and reduce minority carrier lifetimes. This effects, for example, a reduction of solar cell efficiencies [1] and an increase in device leakage currents [2].

There are three main sources of contamination which can lead to electrically active defects: (i) contamination of the bulk silicon crystal due to contaminated feedstock or crystal growth equipment, (ii) contamination of the wafer surface during wafer preparation (e.g. cutting, polishing), and (iii) contamination during device processing. The first two contamination pathways were tackled systematically in the 1980s and 1990s. Contamination levels of metals, which are the main impurities in device processing, are below 10^{11} cm^{-3} in the material used today in microelectronics.[3] Process contamination, however, is still a viable threat in microelectronic device production.

Besides the expensive nearly metal-free silicon used for microelectronics, cheaper and less refined material is often used in the solar cell industry, so-called solar grade silicon.[4–8] In contrast to microelectronic devices, defects in solar cells do not lead to a complete breakdown of their functionality. However, the resulting decrease in efficiency has to be weighed against the reduction in material cost.

Transition metals (TM) are common impurities in the silicon feedstock and are often electrically active.[9–12] Among the TMs, copper, nickel, and cobalt are the three fastest diffusing species in silicon.[11] They diffuse in the interstitial species, but are electrically active in the substitutional species.[12] From a technological point of view, their high diffusivity increases the danger of contamination even at moderate temperatures and also leads to a high likelihood of precipitation. The electrical and structural properties of copper and its interaction with hydrogen have recently been well established.[13–17] Those of cobalt and nickel are discussed in this thesis. In contrast to copper, nickel, and cobalt, titanium is one of the slowest diffusing TMs in silicon.[12] It is electrically active in the interstitial site.[12] While the low diffusivity of titanium makes a contamination, for example during processing, less likely, it is also very hard to remove by annealing processes. For example, annealing a titanium containing silicon sample at 950 °C for nearly 3 h does not fully remove the electrically active titanium (see Chap. 6).

In addition to creating deep levels in the band gap by themselves, the TMs can form complexes with other impurities. A very common and reactive impurity in silicon is hydrogen. Its reactivity is often actively used in industry processes. For example, antireflection coatings on solar cells are used to introduce hydrogen into the top layer of the cell to passivate possible defects.[18] However, hydrogen is also inevitably introduced into the silicon lattice by wet-chemical etching [19, 20] and can even be found in as-grown and prepared wafers [21].

Hydrogen reacts easily with other defects, passivating surface states [22–24] and shallow donors and acceptors [25]. It reacts with the TMs, forming both electrically active and passive TM-H complexes.[26–34] Hydrogen can also activate formerly inactive defects like carbon [21, 35–38] or form other active centers [39, 40], dissolve precipitates [41], and enhance the formation of thermal donors [42, 43]. Furthermore, hydrogen itself has electric levels in the band gap of silicon [44–46]. In high concentrations, it can form extended structures in the lattice like platelets [39, 47–49] or form H_2 molecules [50, 51]. The complexity of possible reactions of hydrogen in the silicon lattice together with its technological use warrant further scientific investigations into the behavior of hydrogen in silicon and its reactions with other defects.

In this work, the properties of the TMs cobalt, nickel, and titanium in silicon, and their interaction with hydrogen are studied by deep level transient spectroscopy (DLTS) and minority carrier transient spectroscopy (MCTS). Investigation of the complexes of these metals with hydrogen started in the 1990s.[29–34] With the access to new techniques, open questions and inconsistencies can now be resolved. A natural starting point for this study is to investigate which hydrogen-related defects are created in metal-free silicon samples after hydrogenation.

This thesis is structured as follows: chapter 2 discusses the theory of deep traps in semiconductors. Chapter 3 contains details about the measurement techniques used, while chapter 4 presents the used samples and preparation techniques. In chapter 5, the results of hydrogenation of metal-free silicon and the influence of the hydrogen isotope on the electrical properties are discussed. The results on cobalt, titanium, and nickel are presented in the chapters 7-8. Each results chapter contains an introduction into the specific topic, the presentation of the results themselves, the discussion of these results, and a short summary. This thesis ends by comparing the studied metals with each other and with those adjacent in the periodic table.

Chapter 2

Theory

This chapter discusses the physics of the phenomena investigated in this thesis. It will start with an analysis of the carrier capture and emission process of deep traps, describing how these processes depend on temperature, carrier concentration, and intrinsic properties of the trap. Next, the time dependence of the carrier capture and emission is considered. These parts follows closely Chap. 7 in P. BLOOD AND J.W. ORTON *The Electrical Characterization of Semiconductors: Majority Carriers and Electron States*.[52] In the third section, the influence of an electric field on the emission process is discussed in dependence of the charge state of the trap. It will outline the basics of the POOLE-FRENKEL theory and of tunneling processes. This chapter ends with a consideration of the penetration behavior of hydrogen during wet-chemical etching and the subsequent formation of hydrogen complexes. A method to determine the number of hydrogen atoms in a defect structure, as proposed by FEKLISOVA AND YARYKIN, is presented.[53]

2.1 Deep traps – capture and emission

The characteristic property of a semiconductor is the separation of two electronic bands, the so-called conduction and valence band. At temperature $T = 0$ K, the valence band is filled with electrons, whereas the conduction band is empty of electrons. The energy region between these bands is called the band gap, in which no allowed electronic states exist. If defects are introduced into the crystal lattice, they can create localized states which lie inside the band gap. These are either called shallow or deep, depending on their distance to the band edges.

Four types of interaction with the bands are possible for any deep (and shallow) level: electron capture from the conduction band, hole capture from the valence band, electron emission into the conduction band, and hole emission into the valence band. These inter-

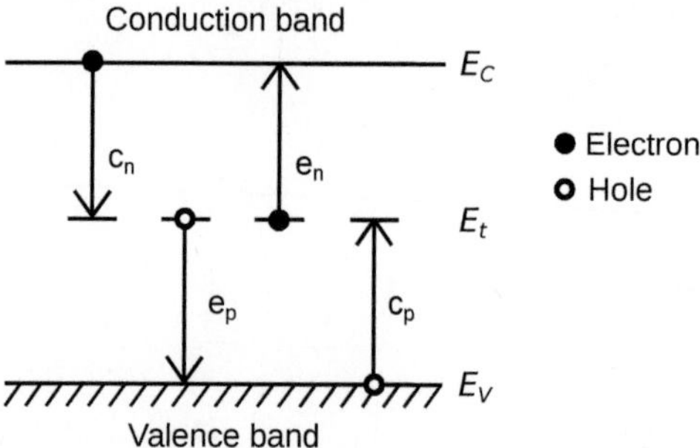

Figure 2.1: Energy diagram of a semiconductor with a deep trap in its band gap. The four possible interactions of the trap with the bands are shown, which are electron/hole capture/emission.

actions are governed by the corresponding capture and emission rates c_n, c_p, e_n, and e_p. Here, n stands for electron processes and p for hole processes. A schematic overview of these processes is given in Fig. 2.1. With these emission and capture rates, the time dependence of the concentration of traps filled with electrons n_t can be written as

$$\frac{\mathrm{d}n_t}{\mathrm{d}t} = (c_n + e_p)(N_t - n_t) - (e_n + c_p)n_t \tag{2.1}$$

with the total trap concentration N_t.

The capture process is characterized by the capture cross section σ, which is an intrinsic property of the defect. Furthermore, it depends on the number of carriers available for capture per unit time. For free carriers, this is the product of the concentration n (p) and the root mean square thermal velocity $\langle v_n \rangle$ $(\langle v_p \rangle)$. The capture rate for electrons is then

$$c_n = \sigma_n \langle v_n \rangle n \tag{2.2a}$$

and for holes

$$c_p = \sigma_p \langle v_p \rangle p. \tag{2.2b}$$

To describe the emission process, the balance of emission and capture processes in thermal equilibrium is considered, that is, Eq. 2.1 is equal to zero. The principle of detailed balance holds, that is, both the emission and capture of electrons and of holes have to be in balance separately. Therefore both

$$e_n n_t = c_n (N_t - n_t) \tag{2.3a}$$

and

$$e_p (N_t - n_t) = c_p n_t \tag{2.3b}$$

have to be fulfilled.

From these equations it follows that in thermal equilibrium the occupancy of the trap is given by

$$\frac{n_t}{N_t} = \frac{c_n}{c_n + e_n} = \frac{e_p}{e_p + c_p}.$$ (2.4)

Another description of the thermal equilibrium occupancy can be obtained by using the Fermi-Dirac distribution, which yields

$$\frac{n_t}{N_t} = \left\{ 1 + \frac{g_0}{g_1} \exp\left(\frac{E_t - E_F}{kT}\right) \right\}^{-1}$$ (2.5)

for a trap with an energy level at E_t, degeneracy g_0 when empty of electrons and g_1 when filled with electrons, and with the Fermi energy E_F of the system. k denotes the Boltzmann constant.

Combining equations 2.4 and 2.5 results in

$$\frac{e_n}{c_n} = \frac{g_0}{g_1} \exp\left(\frac{E_t - E_F}{kT}\right)$$ (2.6)

for the case of electron emission. Since the processes for electrons and holes are decoupled, the equations are from here on derived for electron processes only. Analogue descriptions are valid for the hole processes.

The carrier concentration in a non-degenerate semiconductor can be described by the well-known equation

$$n = N_c \exp\left(-\frac{E_c - E_F}{kT}\right)$$ (2.7)

with the effective density of states in the conduction band N_c. Then, n is substituted in Eq. 2.2a and c_n in Eq. 2.6 to derive the general formula for the emission rate

$$e_n(T) = \sigma_n \langle v_n \rangle \frac{g_0}{g_1} N_c \exp\left(-\frac{E_c - E_t}{kT}\right).$$ (2.8)

To obtain the temperature dependence of the emission process, one substitutes

$$N_c = 2 M_c \left(\frac{2\pi m^* kT}{h^2}\right)^{\frac{3}{2}}$$ (2.9)

and

$$\langle v_n \rangle = \sqrt{\frac{3kT}{m^*}}$$ (2.10)

with M_c being the number of conduction band minima, m^* the effective electron mass, and h the Planck constant. The capture cross section can also be temperature dependent. When the energy the electron gains during capture is dissipated by the emission of multiple phonons, the temperature dependence takes the form

$$\sigma(T) = \sigma_\infty \exp\left(-\frac{\Delta E_\sigma}{kT}\right)$$ (2.11)

with an energy barrier for capture ΔE_σ. For more information about the multiphonon emission process, refer to Refs. [54–57].

Including the aforementioned substitutions, the emission rate can be written as

$$e_n(T) = \gamma T^2 \sigma_{na} \exp\left(-\frac{E_{na}}{kT}\right)$$

(2.12)

with the material constant $\gamma = 2\sqrt{3}M_c(2\pi)^{\frac{3}{2}}k^2 m^* h^{-3}$, the apparent capture cross section $\sigma_{na} = \frac{g_0}{g_1}\sigma_\infty$ and the activation energy $E_{na} = E_c - E_t + \Delta E_\sigma$. For n- and p-type silicon, $\gamma = 7.03 \times 10^{21}\ 1/\mathrm{s\,cm^2\,K^2}$ and $\gamma = 2.64 \times 10^{21}\ 1/\mathrm{s\,cm^2\,K^2}$, respectively.

Equation 2.12 is the final result on the behavior of carrier emission from a deep trap. The two characteristic parameters of a trap follow from this formula, namely the activation energy E_{na} and the apparent capture cross section σ_{na}. Both can be determined by measurements of the emission rate versus temperature.

It should be mentioned that the activation energy E_{na} determined by electrical measurements with variable temperature is, thermodynamically speaking, not exactly equal to the energy level of the trap. This is due to the fact that the size of the band gap is also temperature dependent. The emission of a carrier can be interpreted as the change in chemical potential for the formation of a free carrier and an ionized defect. Then, the activation energy can be described by the change in Gibbs free energy $\Delta G(T)$, which is itself temperature dependent. With the thermodynamical identity

$$\Delta G(T) = \Delta H - T\Delta S$$

(2.13)

the Gibbs free energy can be described by the enthalpy H and the entropy S. Substituting Eq. 2.13 for E_{na} in Eq. 2.12, the emission rate has a temperature dependence of the form

$$e_n(T) \propto T^2 \exp\left(-\frac{\Delta H}{kT}\right).$$

(2.14)

Therefore, the activation energy determined by the electrical measurements is strictly speaking an enthalpy. For more detailed information, see Chap. 8 of Ref. [52] and references therein.

2.2   Deep traps – transients

The time dependence of the occupancy of a trap is given by Eq. 2.1. Solving this differential equation with the initial concentration $n_t = n_t(0)$ at $t = 0$ yields

$$n_t(t) = \frac{a}{a+b}N_t - \left(\frac{a}{a+b}N_t - n_t(0)\right)\exp[-(a+b)t]$$

(2.15)

with the rates increasing the electron population on the deep level

$$a = c_n + e_p \tag{2.16}$$

and those reducing the electron population

$$b = e_n + c_p. \tag{2.17}$$

The steady state at $t = \infty$ is then

$$n_t(\infty) = \frac{a}{a+b} N_t. \tag{2.18}$$

This is a similar result to Eq. 2.4, where the principle of detailed balance was additionally considered.

Usually, one type of carriers (electrons or holes) dominates the system. These carriers are called majority carriers. Since $c \propto n$ and $n_{min} \ll n_{maj}$, $c_{min} \ll c_{maj}$. Then, capture of minority carriers can be neglected. In general, emission of minority carriers can also be neglected, since majority carrier traps are measured in the corresponding half of the band gap and $e \propto \exp(E_c - E_t)$.

Two specific boundary conditions will be analyzed in more detail. The first one corresponds to a pure capture process. In this situation, the initial concentration of filled traps is zero $n_t(0) = 0$. With the considerations from above, Eq. 2.15 becomes

$$n_t(t) = \frac{c_n}{c_n + e_n} N_t \left\{ 1 - \exp[-(c_n + e_n)t] \right\}. \tag{2.19}$$

At the temperature where the emission transients are measured, the capture process is usually much faster than the emission process. Therefore, one can often neglect the emission rate in this equation and thus receives the simplified formula

$$n_t(t) = N_t \left[1 - \exp(-c_n t) \right]. \tag{2.20}$$

The second boundary condition considered is $n_t(0) = N_t$, that is, all traps are filled with electrons. In addition, a situation is assumed in which no free carriers are available for capture and therefore $c_n = 0$. This is for example the case in the depletion region of a Schottky diode. Then, the emission transient can be described as

$$n_t(t) = N_t \exp(-e_n t). \tag{2.21}$$

2.3 Electric field dependence

The emission of carriers from a deep trap can be also influenced by the presence of an electric field. The carrier is trapped in the potential of the defect, the form of which depends on

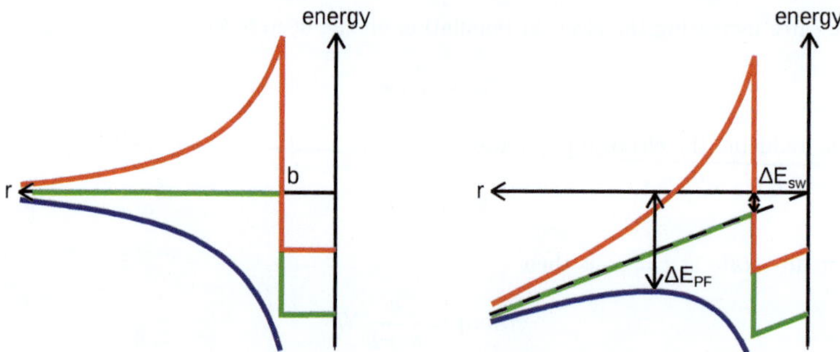

Figure 2.2: Defect potentials in real space for different charge states: attractive (blue), neutral (green), and repulsive (red). The left hand side shows the undisturbed potentials. On the right hand side, the deformation and corresponding barrier lowering by an applied electric field (black dashed line) is shown.

the charge state of the defect. The possible potentials are depicted in Fig. 2.2. If a donor level emits an electron, it is positively charged after the emission of the electron. Then, the potential would be an attractive Coulomb potential (blue line in Fig. 2.2). In the case of a single acceptor which is neutral after the emission, one can approximate the potential by a square well potential (green line in Fig. 2.2). For a double acceptor, the defect potential would be a combination of a square well potential close to the core of the defect with a repulsive Coulomb potential for larger distances (red line in Fig. 2.2).

It should be noted that the charge states associated to the potentials change when one considers hole emission instead of electron emission. Then, the attractive Coulomb potential corresponds to a single acceptor state, the square well potential to a single donor state, and the repulsive Coulomb potential to a double donor state.

In any case, an applied electric field has to be superimposed on the defect potential. This results in the reduction of the potential barrier in the direction of the field (ΔE_{PF} and ΔE_{SW} in Fig. 2.2). The field dependence of this reduction varies between the different possibilities, though.

In the case of the square well potential, the energy barrier is reduced at the edge of the potential well, which is at the distance b. Calculating the effect on the emission rate in three dimensions, one obtains [58]

$$\frac{e_n}{e_{n0}} = \left(\frac{kT}{2eEb} \right) \left[\exp \left(\frac{eEb}{kT} \right) - 1 \right] + \frac{1}{2} \tag{2.22}$$

with the electric field E and the elemental charge e. The emission rate without electric field is denoted with e_{n0}.

For the repulsive Coulomb center, the dependence is about the same, as the barrier lowering also happens at the position b defined by the core square well potential.

It should be noted that the field shift in the square well potential depends strongly on b, a value which is not experimentally accessible and has to be fitted to the data. Values larger than a few nm have to be considered with care, since the lattice constant of silicon is only about 0.54 nm [59].

An attractive Coulomb potential shows a different behavior. The reduction of the barrier leads simultaneously to a shift of the position of its maximum. The mathematical description was first given by FRENKEL for a 1D case and later extended to a 3D model by HARTKE.[58, 60] The effect is usually called the POOLE-FRENKEL effect.

In a 1D case, the barrier lowering leads to a shift of the emission rate of the form

$$\frac{e_n}{e_{n0}} = \exp\left(\frac{\beta\sqrt{E}}{2kT}\right) \tag{2.23}$$

with

$$\beta = \sqrt{\frac{e^3}{\pi\varepsilon\varepsilon_0}}$$

and with the vacuum permittivity ε_0 and the relative permittivity ε.

In the HARTKE model, the effect is calculated for a spherical Coulomb potential. This yields the following equation for the ratio of the emission rates with and without electric field:

$$\frac{e_n}{e_{n0}} = \left(\frac{kT}{\beta\sqrt{E}}\right)^2 \left\{1 + \left[\left(\frac{\beta\sqrt{E}}{kT}\right) - 1\right]\exp\left(\frac{\beta\sqrt{E}}{kT}\right)\right\} + \frac{1}{2} \tag{2.24}$$

The difference between the 1D and the 3D model is more pronounced at low temperatures. At higher temperatures, the models yield nearly identical slopes. It depends on the defect which of the two models describes the real situation more accurately. Real potential distributions of defects in the host lattice are usually more complicated than a simple sphere (see, e.g., Ref. [28]). Therefore, the emission rate shift in an electric field may follow the HARTKE model for nearly spherical distributions, whereas strongly anisotropic potentials may be better described by the 1D model.

The shift of the emission rate in the electric field becomes weaker for higher temperatures. This weakening makes the experimental determination of a POOLE-FRENKEL effect at higher temperatures difficult. For example, at 280 K the emission rate shifts only by a factor of ~ 1.6 between 2500 V/cm and 15000 V/cm, which are typical fields in the depletion region of a Schottky diode for $n \approx 10^{15} \text{cm}^{-3}$.

In contrast to the square well model, the POOLE-FRENKEL models yield a fixed field dependence for a given temperature. No arbitrary fit parameters are used in the comparison with experiments. It should be noted, though, that sometimes the experimentally observed shifts of the emission rate show a weaker field dependence than expected from the POOLE-FRENKEL model.[61] However, they still follow the $\sqrt{E}$ proportionality.

Finally, for sufficiently high electric fields, tunneling processes start to contribute to an enhancement of the emission rate. The change in the potential induced by the electric field changes the tunneling parameters. Following the model of KARPUS AND PEREL' [62], the emission rate is enhanced in the form

$$e_n(E) \propto \exp\left[\frac{(eE)^2\tau^3}{3m^*\hbar}\right].$$

(2.25)

Here, τ is the tunneling time.

Since the emission rate shift for tunneling is $\propto E^2$ in contrast to the $\sqrt{E}$ dependence of the POOLE-FRENKEL effect, a differentiation of these two processes is possible. Also, sometimes a transition from the $\sqrt{E}$ to the E^2 dependence is observed at sufficiently high electric fields.

2.4 Depth distribution of hydrogen complexes

In the case of an inhomogeneous hydrogen concentration, the concentration of hydrogen-containing complexes varies accordingly. But the exact concentration profiles depend also on the number of hydrogen atoms involved in the complex. This dependence can be used to obtain information about their microscopic structure. For the case of hydrogenation by wet-chemical etching, the concentration profiles of hydrogen complexes in dependence of the number of hydrogen atoms involved were calculated by FEKLISOVA AND YARYKIN [53]. Since the results of their calculations are used extensively throughout this thesis, they are briefly presented here.

During wet-chemical etching, the number of mobile hydrogen atoms $[\mathrm{H}(x)]$, with the distance to the surface x, can be described by the differential equation

$$\frac{\partial[\mathrm{H}]}{\partial t} = D\frac{\partial^2[\mathrm{H}]}{\partial x^2} + V\frac{\partial[\mathrm{H}]}{\partial x} - \frac{[\mathrm{H}]}{\tau}$$

(2.26)

with the diffusion coefficient D and lifetime τ of mobile hydrogen.[53] The second term on the right hand side of the equation accounts for a moving surface during the etch process with etching velocity V. The solution of this equation is the simple exponential behavior

$$[\mathrm{H}] = \mathrm{H}_0 \exp(-x/L)$$

(2.27)

with the penetration depth L.

Assuming that a given complex AH_i of the defect A with i hydrogen atoms is created by the reaction $AH_{i-1} + H = AH_i$, the formation of AH_i is governed by

$$\frac{\partial[AH_i]}{\partial t} = V\frac{\partial[AH_i]}{\partial x} + 4\pi D(r_{i-1}[AH_{i-1}] - r_i[AH_i])[H] \qquad (2.28)$$

with the radius of hydrogen capture r_i for the defect AH_i.

Considering the concentration in sufficiently large depth where $[AH_{i-1}] \gg [AH_i]$, the second term in the brackets can be neglected. Then, one obtains the quasistationary solution for $i = 1$

$$[AH_1] = 4\pi D r_0 H_0 \frac{L}{V} \exp(-x/L). \qquad (2.29)$$

For a free choice of i, the solution takes the form

$$[AH_i] \propto \exp\left(ix/L\right). \qquad (2.30)$$

With this proportionality, the ratio of hydrogen atoms involved in different complexes can be determined from the ratio of the slopes in a semilogarithmic plot of concentration versus depth. This allows a direct determination of the microscopic structure of hydrogen-related complexes.

As mentioned above, this solution is valid for hydrogen introduction by wet-chemical etching. This is expressed in the second term in the differential equation (see Eq. 2.26). Other ways of creating mobile hydrogen in the sample, such as plasma hydrogenation or annealing of a hydrogenated sample, would require a separate analysis.

Chapter 3

Measurement techniques

Using the theoretical basis from the previous chapter, the measurement techniques used in this work are described in this chapter. The first two sections cover the basic principles of deep level transient spectroscopy (DLTS) and Laplace DLTS. A section about the determination of a concentration depth profile follows. The chapter ends with an introduction of minority carrier transient spectroscopy (MCTS).

3.1   Deep level transient spectroscopy

DLTS is a technique to measure deep states in the band gap of a semiconductor. It was introduced in 1974 by D. Lang [63] The basic operating principle of DLTS is the measurement of the depletion capacitance in a diode structure. This capacitance results from the presence of charged atoms in the depletion region, mostly those of the main dopant. However, also defects with deep states are charged and contribute to the capacitance, albeit usually in a much smaller way. To discriminate the deep states from the dopant and from each other, their different transient behavior is exploited.

To measure these transients, a diode is kept in reverse bias V_R. Then, a so-called filling pulse collapses the depletion region partially or totally by applying a smaller reverse bias for the duration of the pulse width. During this time, the traps are refilled with majority carriers, a process which is governed by Eq. 2.19. After returning to the reverse bias, the traps re-emit their carriers if the Fermi level has crossed the deep level position. This is depicted in Fig. 3.1, where x denotes the depth under the contact. As this emission changes the total charge in the depletion region, the capacitance changes accordingly. Therefore, the emission process can be monitored in the transient behavior of the capacitance.

The actual measurement is performed by continually applying pulses and measuring the capacitance transients while changing the temperature of the sample. The capacitance tran-

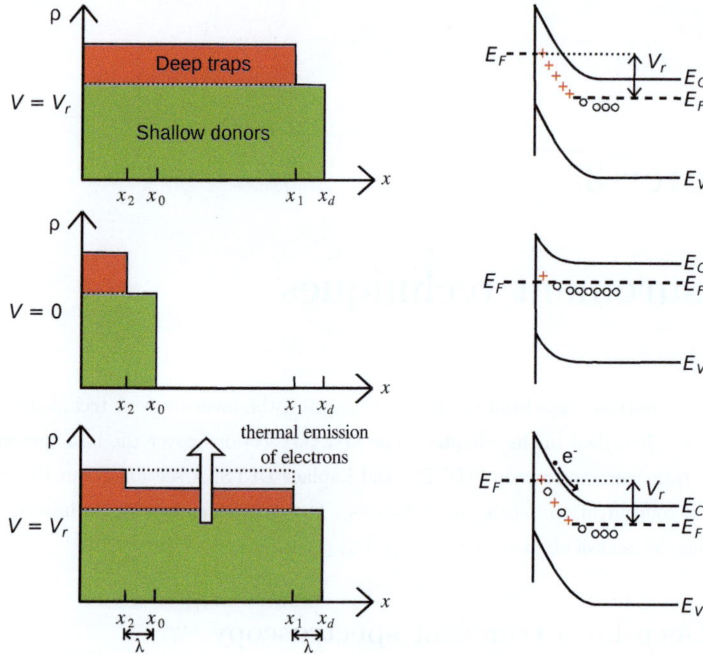

Figure 3.1: Left hand side: behavior of the charge density ρ in the depletion region initially, at the end of the filling pulse, and during measurement (from top to bottom). The right hand side shows the corresponding band bending with the charged traps.

sient is analyzed by a lock-in amplifier, which gives maximum output if the capacitance transient time constant matches with the set lock-in frequency. As the emission rate depends strongly on temperature (see Eq. 2.12), peaks will appear at the temperatures corresponding to the energy levels of the defects. This is illustrated in Fig. 3.2.

In this work, the DLTS peaks are labeled with either E or H for electron and hole traps, respectively. The letter is followed by the temperature at which the peak maximum is observed with an emission rate of $47\,\mathrm{s}^{-1}$. To distinguish similar peaks which arise from different defects, subscripts denote the essential defect component (e.g. Ti for all titanium-related defects). Where applicable, data taken in n-type samples is shown in blue, whereas data taken in p-type samples is shown in red. This color code has to be dropped for concentration depth profiles due to the amount of data displayed there.

From the amplitude of the capacitance change, one can calculate the trap concentration. Assuming a total collapse of the depletion region and completely filled traps after the filling pulse, the simple expression for the capacitance change is [52]

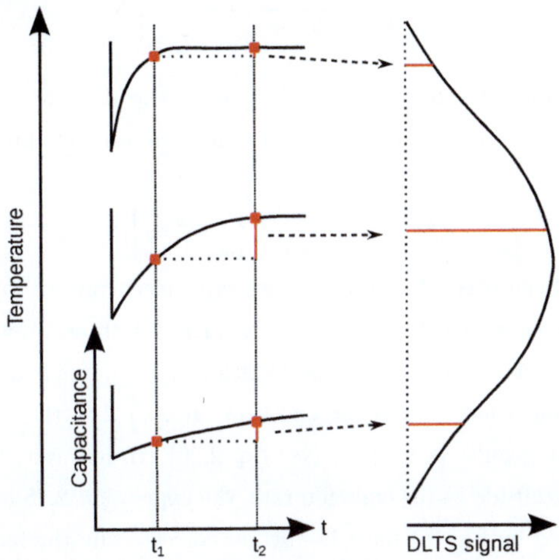

Figure 3.2: Left hand side: emission transients at different temperatures. The transients are sampled with a fixed rate window (t_1 and t_2). The capacitance difference is the DLTS signal. If the emission rate matches the rate window, the signal becomes maximal and a peak appears in the DLTS spectrum (right hand side).

$$\frac{\Delta C(t)}{C(\infty)} = -\frac{N_t}{2N_d} \exp(-e_n t) \tag{3.1}$$

with N_t the total trap concentration, N_d the dopant concentration, ΔC the change in capacitance after the filling pulse, and $C(\infty)$ the saturation capacitance. This expression assumes a constant depletion width during the emission process, as usually $N_t \ll N_d$. Therefore, for a total reduction of the depletion region to $x = 0$, the total trap concentration is

$$N_t = -\frac{2N_d \Delta C(0)}{C(\infty)}. \tag{3.2}$$

Usually, the width of the depletion region is not reduced to zero, but only to the depth corresponding to the filling pulse voltage x_0. That leads to a reduced volume from which the measured signal stems. Furthermore, the depth scale corresponding to the deep traps is not the same as for shallow dopants. The occupancy of the traps is governed by the position of the Fermi level. For deep traps, due to the band bending in a diode structure, the crossover of the Fermi level with the trap level is located closer towards the diode interface than the calculated depletion width. This difference is called the lambda layer and can be calculated as [52]

$$\lambda = \sqrt{\frac{2\varepsilon\varepsilon_0}{e^2 N_d}(E_F - E_t)}. \tag{3.3}$$

With the depletion width during reverse bias x_d and the effective widths $x_1 = x_d - \lambda$ and $x_2 = x_0 - \lambda$, one can then obtain the total trap concentration including the reduced volume by [52]

$$N_t = \frac{2N_d \Delta C(0)}{C(\infty)} \left\{ \frac{x_d^2}{x_1^2 - x_2^2} \right\}. \tag{3.4}$$

Equation 3.4 has been used to determine the trap concentrations in this work. However, there are two special cases which have to be considered with additional care. The first is that of a slow capture rate in respect to the emission rate. So far it has been assumed that the capture rate is much faster than the emission rate, so that the concentration of filled traps n_t after the filling pulse is $n_t = N_t$ (see Eq. 2.20). If, however, the capture rate is of the same order of magnitude as the emission rate, the expression 2.18 has to be used and the concentration obtained by Eq. 3.4 must be further corrected by the factor $(c + e)/c$. In the second special case, the concentration of deep traps is of the same order of magnitude as that of the shallow dopants. In this case, the assumption that the depletion width is constant during filling and during emission is no longer valid, as the charge change due to deep traps is not negligible against that of the shallow dopants. This leads to non-exponential capture and emission transients and a voltage dependence of the capture rate.[64] Also, care has to be taken during the evaluation of the trap concentration to use correct values for the depletion width.

3.2   Laplace DLTS

A drawback of the conventional DLTS using a lock-in amplifier is its low energy resolution. Due to quite broad peaks created by the analogue filtering, defects with similar activation energies are difficult to separate. Generally, emission rates of two defects have to differ by a factor of ~ 12 to ~ 15 to be distinguishable by conventional DLTS. To overcome this limitation, the Laplace DLTS technique is used.[65] Here, the capacitance transient is measured at a constant temperature. By multiple measurements it is possible to statistically reduce the noise level. Then, exponential functions are fitted to the measured transient via a reverse Laplace transformation.

The algorithms used attempt to find the least number of peaks to consistently fit the experimental data.[65] Also, peaks with amplitudes smaller than the noise level are removed from the solution. This requires a good signal to noise ratio (SNR) to separate peaks with similar emission rates. As was demonstrated in Ref. [65], a SNR of about 300 is required to separate two peaks whose emission rates differ by a factor of two. The requirements rise if

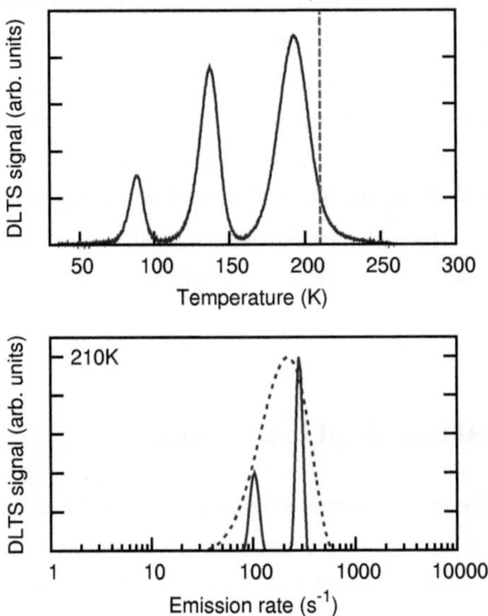

Figure 3.3: Illustration of the resolution of Laplace DLTS. The top graph shows a DLTS spectrum of cobalt in silicon. The bottom graph shows normalized Laplace DLTS data taken at 210K (marked in red in the DLTS spectrum) calculated with the same algorithm. The dotted line has a SNR of ~ 28, the full-drawn line has a SNR of ~ 170. For sufficiently high SNR, the one DLTS peak can be separated into two peaks in Laplace DLTS with an emission rate ratio of about 1:3.

the peaks lie even closer together or if one peak has a significantly smaller amplitude than the other one.

An example of a Laplace DLTS measurement is presented in Fig. 3.3. A DLTS spectrum of cobalt-doped silicon and two normalized Laplace DLTS measurements taken at 210K with the same algorithm and fitting parameters are shown. In the DLTS spectrum, only one peak is observed at this temperature. In Laplace DLTS, a SNR of ~ 28 results in one broad peak (dotted line). If one increases the measurement time to obtain a SNR of ~ 170, it is possible to separate two components. The emission rates are separated by a factor of ~ 3.2.

Variation of the pulse parameters allows the determination of several trap properties: (i) The activation energy can be determined from the emission rates at different temperatures. (ii) The concentration of the defect can be calculated from the peak amplitude (see Eq. 3.4). (iii) A concentration depth profile can be taken by varying both the filling pulse voltage and the reverse bias. To obtain precise concentration data, two measurements with the same reverse bias and different pulse voltages are subtracted from each other. With this so-called double DLTS, effects of the Debye-tail at the depletion region edge are diminished.[66] (iv) The capture cross section can be measured by changing the width of the filling pulse. The dependence of the peak amplitude on the filling pulse width also allows one to distinguish between point and extended defects. Point defects show an exponential dependence on the filling pulse width, while extended defects exhibit a logarithmic dependence.[67] (v) Finally, effects of the electric field on the emission rate can be observed by changing the applied voltage.

3.3   Concentration depth profiling

To obtain a concentration depth profile, several measurements are performed in which the probed region is shifted for each measurement. The corresponding mean depth can be calculated from the CV-profile for each data point. The DLTS amplitude yields the defect concentration.

As was mentioned above, the crossing point of the defect level with the Fermi level, which determines the depth unto which a defect is charged, is shifted with respect to the depletion region edge by the lambda layer (see Eq. 3.3). This lambda layer has to be subtracted from the depth calculated from the CV-profile for a correct concentration depth profile.

Measuring the depth profile with a single filling pulse leads to several complications. First of all, electrons leaking into the depletion region (the Debye tail) lead to a nonzero capture rate close to the depletion region edge, and the observed emission transients are modified accordingly (see Eq. 2.15). In addition, contact and interface effects can influence the DLTS signal and thus the calculated concentration.[66]

In order to avoid these problems, two different filling pulses are applied and the transients subtracted, as proposed by LEFEVRE AND SCHULZ.[66] The subtraction eliminates constant effects which are independent of the filling pulse. In addition, the effective probed region is located far from the depletion edge under reverse bias. Therefore, effects from the Debye tail are also eliminated.

The electric field in the depletion region can influence the emission rate of the observed defect (see Sec. 2.3). In order to avoid possible effects on the measured concentration, the reverse bias and the pulse voltages are selected in such a way that the electric field in the probed region is about the same for all measurements of one profile. The maximum reverse bias applied in the concentration depth profiles in this work was $V_R = -10$ V.

3.4 Minority carrier transient spectroscopy

Schottky diodes are unipolar devices, that is, only one type of carriers is active. DLTS on Schottky diodes can therefore only manipulate the majority carrier concentration. In order to measure minority carrier traps in the same sample, it is possible to introduce minority carriers into the depletion region using light pulses. This technique is called minority carrier transient spectroscopy (MCTS).

For MCTS measurements, a Schottky diode is kept under reverse bias. Electron-hole pairs are generated by illumination with a wavelength whose energy is greater than the band gap. For silicon, the band gap is 1.11 eV at room temperature, therefore the wavelength should be shorter than $\sim$ 1130 nm. The minority carriers are pulled into the depletion region, while the majority carriers are expelled. The minority carriers will fill the traps in the depletion region until a steady state is reached or the illumination is turned off. After the illumination ends, the traps emit the minority carriers, which can be observed as capacitance transients. This process is shown in Fig. 3.4.

The samples are illuminated by shining a laser diode either onto the front or onto the backside of the sample. Illumination of the front side has the disadvantage of creating many majority carriers in the depletion region. This generates majority carrier peaks in the spectrum and can even dominate the spectrum, so that a standard DLTS spectrum is reproduced. Also, interface states between the semiconductor and the contact metal will be excited and contribute to the background.

Illumination from the backside avoids these problems and is therefore desirable. The limiting factor here is the diffusion length of the minority carriers, which should be larger than the sample thickness. In the case of silicon, which is an indirect band gap semiconductor, it is also advantageous to use light with an energy just above the band gap. Due to the low absorption coefficient at such a wavelength, the penetration depth is larger and the minority carriers are generated closer to the depletion region.

As the amount of minority carriers reaching the depletion region depends strongly on a number of parameters, such as light intensity, laser pulse width, diffusion length of the sample, sample thickness etc., a quantitative analysis of the defect concentrations by MCTS is difficult. However, spectra of the same sample taken under different conditions can yield qualitative information about relative concentrations of the observed defects.

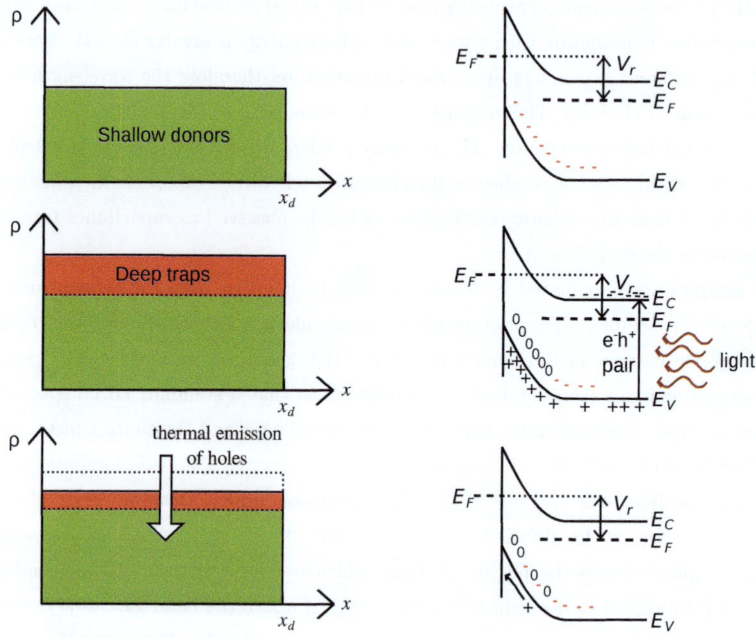

Figure 3.4: Left hand side: behavior of the charge density ρ in the depletion region initially, at the end of the laser pulse, and during measurement (from top to bottom). The right hand side shows the corresponding band bending with the charged traps.

Chapter 4

Experimental details

4.1 Samples and preparation

For the investigation of the transition metals in silicon, both n- and p-type Floating-Zone (FZ)[1] and Czochralski (CZ)[2] silicon samples were used. Into these samples, the metals had been added during crystal growth.

In addition, FZ n- and p-type silicon[3] was used in which vacancies were created by high temperature annealing in nitrogen gas, which stabilizes the vacancies.[68–70] Since this also introduces nitrogen into these samples, they are referred to as nitrogen-doped samples in Chap. 8. Into these samples, nickel was in-diffused at 700 °C which reacts with the vacancies to form substitutional nickel. The in-diffused nickel extended over the entire wafer thickness, with only small concentration changes over length scales in the order of 100 µm.[71] Control samples without in-diffused nickel were also available.

For the investigation of hydrogen in metal-free n-type silicon, both FZ and CZ samples with different doping concentrations were analyzed. In all samples, phosphorous was used for n-type doping and boron for p-type doping. An overview of all sample types used with their corresponding doping concentrations is given in Table 4.1.

Schottky contacts were created by resistive evaporation of gold (n-type) or aluminum (p-type) under vacuum conditions onto the samples. The sample surface was prepared prior to contact evaporation either by cleaving, polishing, or wet-chemical etching. Etching was performed by a mixture of $HF:HNO_3:CH_3COOH$ with a ratio of 3:5:3 (CP4A). Natural surface oxide was removed by an HF-dip of several seconds. Etching by CP4A is always accompanied by an introduction of hydrogen into the surface layer.[19, 20] When this hydrogen

---

[1]provided by Leibniz-Institut für Kristallzüchtung, Max-Born-Str.2, 12489 Berlin
[2]provided by Siltronic AG, Hans-Seidel-Platz 4, 81737 München
[3]provided by P. Saring, Georg-August-Universität Göttingen

| | doping concentration (cm^{-3}) | | | | | |
| | FZ samples | | CZ samples | | diffusion samples | |
	n-type	p-type	n-type	p-type	n-type	p-type
Co	9×10^{13}	-	-	1×10^{15}	-	-
Ni	5×10^{13}	3×10^{13}	-	1×10^{15}	2×10^{14}	2×10^{14}
Ti	1.7×10^{15}	1×10^{14}	1.5×10^{15}	1×10^{15}	-	-
clean	3×10^{13}-4×10^{16}	-	1-6×10^{15}	-	-	-

Table 4.1: Overview of the used samples with their corresponding doping concentrations.

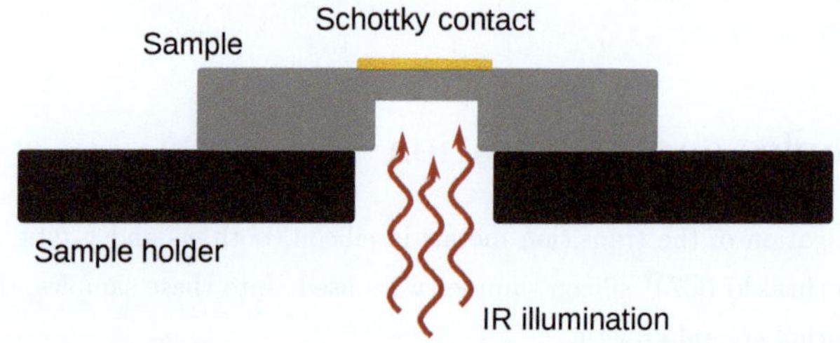

Figure 4.1: Schematic cross-section of a sample prepared for MCTS with backside illumination mounted in the sample holder. Underneath the Schottky contact, the sample is etched to a thickness of around 200 µm. The rest of the sample retains its original thickness for mechanical stability. Illumination is provided from the back through a hole in the sample holder.

introduction was unwanted, samples were prepared by cleaving or polishing. Ohmic contacts were prepared by rubbing an eutectic InGa alloy onto the backside of the sample.

Samples used for MCTS with backside illumination were partially etched from the back right underneath the contact. The etching was done in such a way that the sample thickness under the contact is in the order of $\sim$ 200 µm, while the surrounding material retains its original thickness of $\sim$ 500 µm. In this way, the mechanical stability of the samples is retained while at the same time the diffusion pathway for the minority carriers is reduced. A schematic cross-section of a prepared MCTS samples is presented in Fig. 4.1.

Hydrogen introduction into the samples was achieved by wet-chemical etching. Additionally, a remote DC-hydrogen plasma was available for sample preparation. A schematic of the plasma setup used is given in Fig. 4.2. The plasma is ignited by a DC high voltage (1kV) at a distance of about 10 cm from the sample. Hydrogen ions from the plasma are pulled towards the sample by an applied bias. The sample stage can be heated up to 400 °C. Besides hydrogen, deuterium and a deuterium-hydrogen (50%-50%) gas mixture can be used for the plasma.

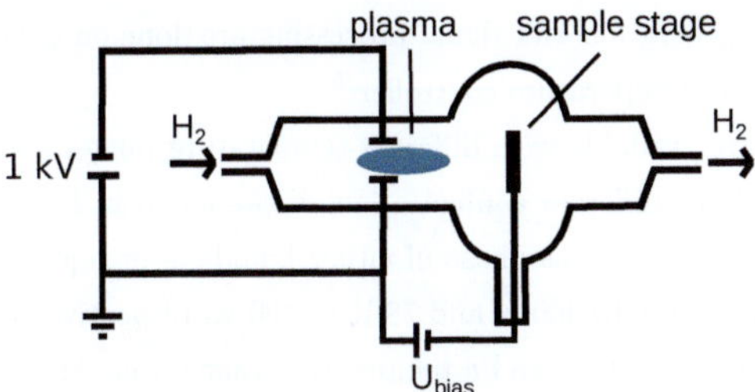

Figure 4.2: Schematic of the setup used for plasma hydrogenation. The plasma is ignited by a DC high voltage. Hydrogen ions are pulled towards the sample by an additional applied bias. The sample stage can be heated. Besides hydrogen, deuterium and a hydrogen-deuterium gas mixture can be used.

In order to determine the temperature dependence of the defects, the samples were annealed at different temperatures. Annealing up to 400 K was often performed with the sample mounted in the sample holder. Temperatures higher than 400 K lead to a degradation of the ohmic contact and the silver paste used to glue the sample to the sample holder.

Annealing up to 300 °C was mostly performed in air. For higher temperatures, a horizontal and a vertical quartz tube furnace were available. In the horizontal furnace, the samples were annealed under an argon flow in an open quartz boat. In the vertical furnace, sealed quartz ampules with helium atmosphere were used. Quenching times for the annealing in air and the open quartz boat are in the order of tens of seconds. In the vertical furnace, the quartz ampule is dropped directly into a water basin, resulting in quenching times below a second. Annealing experiments at temperatures above 150 °C were always performed without Schottky contacts.

4.2   Instrumentation

Conventional DLTS measurements are performed using stand-alone pulse generator[4], capacitance bridge[5], lock-in amplifier[6], and temperature controller[7] controlled via LabView. For

---

[4] Agilent 81110 A
[5] Boonton 72B
[6] EG&G 7260 DSP
[7] LakeShore DRC 91CA

Laplace DLTS, pulse generation and signal processing are done on-board[8] with an external capacitance bridge[9] and temperature controller[10].

Several cryostats are available with different temperature ranges and special features. A schematic overview of the different configurations is presented in Fig. 4.3. Two home-built sample holders are used in the gas phase of either liquid He or liquid N_2. Their accessible temperature ranges are 30 K to 400 K and 78 K to 400 K, respectively. A cold-head sample holder cooled by flowing liquid N_2 and a temperature range from 90 K to 400 K contains an in-built IR laser diode with a laser wavelength of 780 nm and can be used for frontside and backside illumination MCTS measurements. Finally, a closed-cycle Janis cryostat covers a temperature range from 10 K to 450 K with windows for backside illumination MCTS measurements.

---

[8]National Instruments 6251 data acquisition board
[9]Boonton 7200
[10]LakeShore 331

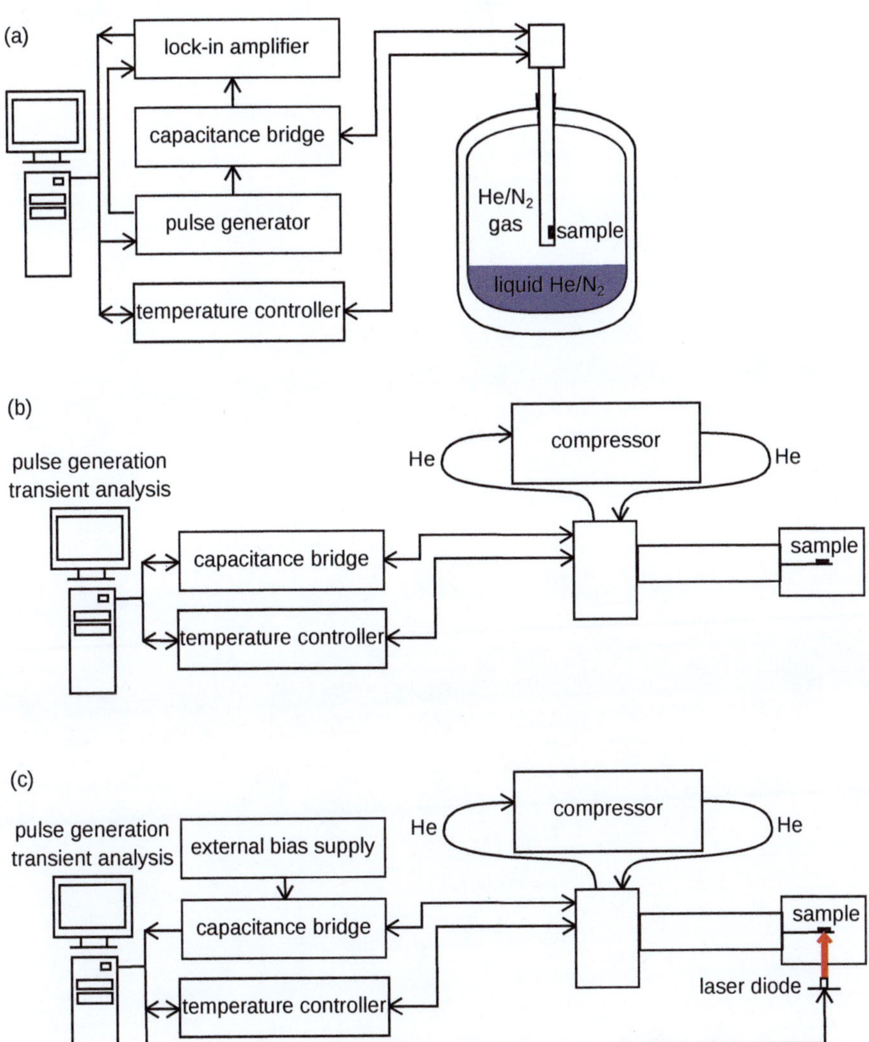

Figure 4.3: Schemes of the setups used for measurement. In (a) the conventional DLTS setup with dewar cooling is shown. The closed cycle cryostat is shown in (b) used for Laplace DLTS and in (c) for MCTS.

Chapter 5

Hydrogenation of metal-free silicon

Before studying the interaction of TMs with hydrogen, it is important to understand which defects appear in metal-free silicon after hydrogenation. To that end, the hydrogenation processes used on the TM-containing samples were also applied to metal-free silicon samples. This chapter presents the results of the investigations of these samples.

5.1 Introduction

Previous studies on the effect of hydrogen have shown the appearance of different hydrogen-related defects in n- and p-type silicon.[21, 35–37, 39, 40, 72, 73] The effect of plasma hydrogenation on silicon was studied by JOHNSON $et\ al.$[39] They reported on the formation of planar microdefects called platelets in the first 0.1 µm below the surface. In addition, two deep levels in the band gap at E_C - 0.06 eV (E45$_C$) and E_C - 0.51 eV (E260$_C$) were observed deeper below the surface.[39] Similar levels were also reported in Refs. [40, 72] in plasma hydrogenated silicon. Apart from the involvement of hydrogen in these defects, their origin remained unclear.

Two other levels E1 (E65$_C$) and E2 (E75$_C$) lying at E_C - 0.11 eV and at E_C - 0.13 eV were reported by YONETA $et\ al.$ and others in wet-chemically etched silicon.[21, 35] The concentration of these defects was correlated with the concentrations of carbon and oxygen. Therefore, they were tentatively attributed to two different COH complexes.[35]

A third level E3 (E90$_C$), lying at about E_C - 0.16 eV, was also observed by these authors.[21, 35] Its properties are similar to those of a defect observed by ENDRÖS $et\ al.$, who assigned this defect to the donor level of a CH-complex.[37] ENDRÖS $et\ al.$ used reverse bias annealing in hydrogenated samples to create this defect. Other studies observed a similar CH-complex after wet chemical etching at room temperature and after low temperature proton implantation and subsequent annealing above 225 K, which introduces also a second level at $E_V + 0.29$ eV.[36] However, a different behavior of the emission rate under an applied electric field was observed for the complexes created with or without reverse bias annealing. This discrepancy could recently be resolved by STÜBNER *et al.*, who could show that these are in fact two different configurations of CH with different charge states.[73]

The low temperature proton implantation study mentioned above additionally observed a donor level at E_C - 0.22 eV.[36] It was attributed to a further configuration of the CH-complex. It is instable for temperatures above 225 K and transforms into E3.

Theory calculates at least five different configurations of the CH complex.[36] The configuration proposed to be most stable has the hydrogen atom located in the bond-centered position between the carbon and a neighboring silicon atom (CH_{1BC}). An only 0.2 eV higher formation energy was calculated for CH_{2BC} in the positive charge state. In this configuration, the hydrogen atom is located in the bond between the nearest and second nearest neighbor silicon atoms. A third configuration is CH_{1AB}, where hydrogen sits in the anti-bonding position bound to carbon. Here, the formation energy is in the negative charge state only 0.2 eV higher than that of CH_{1BC}. Also possible is the position in the interstitial site next to the carbon atom, CH_{1TD}. The anti-bonding configuration with the hydrogen bound to the silicon atom (CH_{2AB}) was found to have significantly higher formation energies in all charge states than all other configurations considered.

ANDERSEN *et al.* attributed the donor level at E_C - 0.22 eV after low temperature proton implantation to CH_{2BC} and the CH complex reported in Refs. [21, 35–37] at E_C - 0.16 eV to CH_{1BC}. The assignment of the CH complex observed directly after etching to CH_{1BC} was supported in Ref. [73]. The complex created after reverse bias annealing was attributed to a different configuration CH_B.[73]

No experimental evidence has been reported so far on the existence of the other configurations of CH. With the results presented in this chapter, $E45_C$ and $E260_C$ can be attributed to two charge states of the same carbon-hydrogen configuration.

5.2   Results

DLTS   Figure 5.1 shows the DLTS spectra of metal-free n-type FZ silicon with a doping concentration of 1×10^{15} cm^3 after hydrogen introduction by etching and by a DC hydrogen plasma. The hydrogen plasma treatment was performed at different sample temperatures, ranging from 50 °C to 200 °C in steps of 50 °C. Note the different scaling factors of the spectra. Three peaks are visible in these spectra, $E45_C$, $E90_C$, and $E260_C$. $E90_C$ was only observed in the etched FZ samples. The intensities of $E45_C$ and $E260_C$ depend strongly on the sample temperature during the plasma treatment.

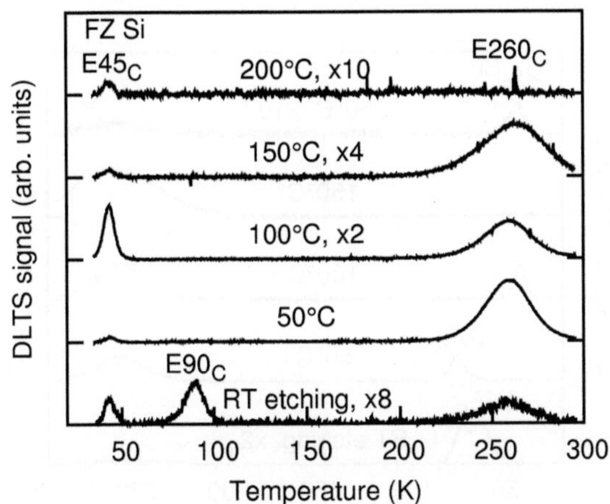

Figure 5.1: DLTS spectra of metal-free FZ silicon after wet-chemical etching and after DC hydrogen plasma at the indicated temperatures. The spectra were measured with $V_R = -2$ V, $V_P = 0$ V, a filling pulse width of 1 ms, and a rate window of 47 s^{-1}.

In Fig. 5.2, the DLTS spectra of metal-free n-type CZ silicon with a doping concentration of 5×10^{14} cm^3 are shown after the same treatment as the FZ samples. In contrast to the FZ samples, the etched CZ sample shows two peaks E65$_C$ and E75$_C$ instead of E90$_C$. In the plasma hydrogenated CZ silicon samples, E45$_C$ and E260$_C$ are again observed with a similar temperature dependence as in FZ silicon.

Arrhenius plot   The activation energies and apparent capture cross sections of the observed peaks were determined from their Arrhenius plots shown in Fig. 5.3. The results are given in Table 5.1.

label	E_{na} (eV)	σ_{na} (cm^2)	assignment
E260$_C$	- 0.50	$(3.3 \pm 1.8) \times 10^{-15}$	CH$_{1AB}$ (-/0) [this work]
E90$_C$	- 0.15	4.7×10^{-16}	CH$_{1BC}$ (-/0) [36, 73]
E75$_C$	- 0.13	7.6×10^{-16}	COH(0/+) [21, 35]
E65$_C$	- 0.12	3.7×10^{-15}	COH(0/+) [21, 35]
E45$_C$	- 0.07	2.3×10^{-16}	CH$_{1AB}$(- -/-) [this work]

Table 5.1: Activation energies, apparent capture cross sections, and charge states of the peaks observed in hydrogenated metal-free n-type silicon.

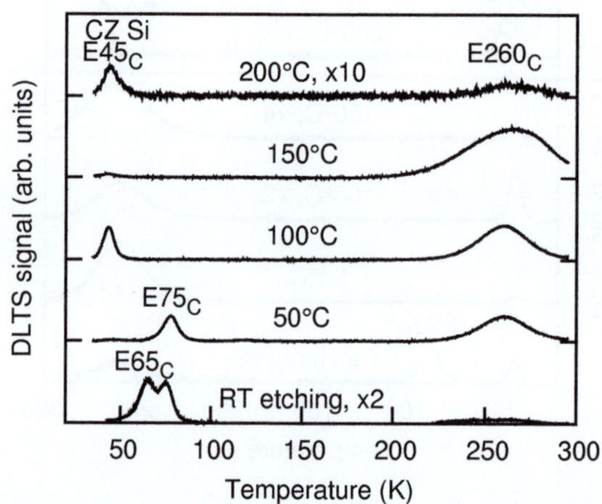

Figure 5.2: DLTS spectra of metal-free CZ silicon after wet-chemical etching and after DC hydrogen plasma at the indicated temperatures. The spectra were measured with $V_R = -2$ V, $V_P = 0$ V, a filling pulse width of 1 ms, and a rate window of 47 s^{-1}.

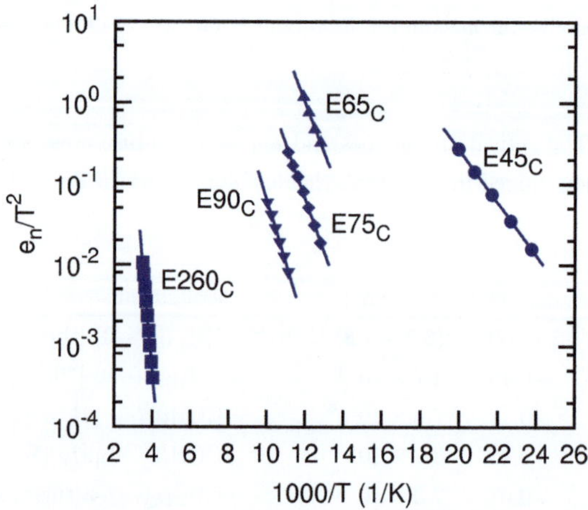

Figure 5.3: Arrhenius plots of the peaks observed in hydrogenated metal-free n-type silicon.

Electric field dependence The emission rates of the peaks $E45_C$, $E65_C$, and $E75_C$ are shown in Figs. 5.4 and 5.5 as a function of the square root of the electric field. Both $E65_C$ and $E75_C$ show good agreement between experiment and theory and can therefore be assigned to donor levels (see Fig. 5.4). In contrast, the dependence of $E45_C$ on the electric field is significantly lower than expected from both the 1D-POOLE-FRENKEL model (dotted line) and the HARTKE model (solid line) (see Fig. 5.5). However, a fit with the square-well potential (dash-dotted line) yields good agreement with a well-radius of $b = 4.35$ nm.

The emission rates of $E90_C$ and $E260_C$ do not shift in an applied electric field. Both defects can therefore be ascribed to acceptor levels.

Depth profiles Figure 5.6 shows the concentration depth profiles of $E260_C$ and $E45_C$ in FZ silicon for two different plasma conditions. The concentrations of $E45_C$ and $E260_C$ are identical in both cases.

The coincidence of the concentration profiles is not apparent from the DLTS spectra presented in Figs. 5.1 and 5.2. Instead, different peak intensities of $E45_C$ and $E260_C$ are observed. However, $E45_C$ and $E260_C$ have quite different activation energies. The resulting difference in the lambda layer of the two levels shifts the probed region during the measurement by 0.5 µm in these samples (see Sec. 3.1). Taking into account the non-uniform depth distribution of $E45_C$ and $E260_C$ (see Figs. 5.6 and 5.7), this shift of the probed region explains the different DLTS peak amplitudes of $E45_C$ and $E260_C$.

Concentration depth profiles of $E260_C$ for different plasma treatment temperatures are presented in Fig. 5.7. For simplicity, the profiles of $E45_C$ are omitted. With increasing temperature of the plasma treatment, hydrogen diffuses further into the sample. This leads to a shift of the concentration maximum deeper into the sample and a broader defect distribution.

The concentration depth profiles of $E45_C$ and $E260_C$ measured in a FZ sample with a doping concentration of 3×10^{13} cm^3 after wet-chemical etching are presented in Fig. 5.8, together with the profile of $E90_C$. The concentrations of $E45_C$ and $E260_C$ coincide and are a factor of 10 smaller than the concentration of $E90_C$. Using the slopes of the concentration profiles to compare the number of hydrogen atoms involved in each complex, as presented in Sec. 2.4, one obtains equal slopes for all three levels. This shows that the same number of hydrogen atoms are involved in these defects.

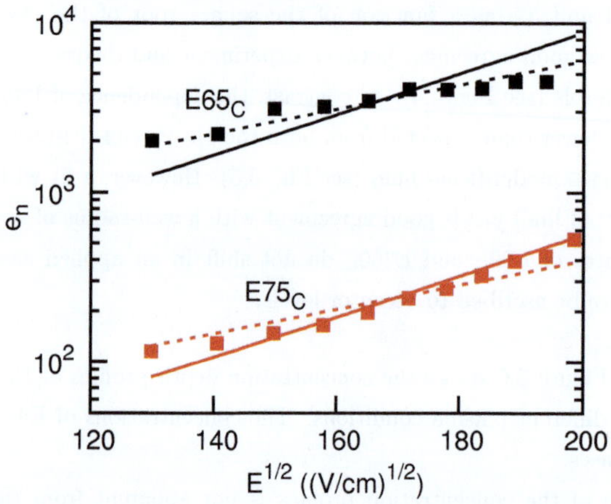

Figure 5.4: Electric field dependence of E65$_C$ and E75$_C$. The lines shown the theoretically expected behavior in the 1D-Poole-Frenkel model (dotted line) and in the Hartke model (solid line).

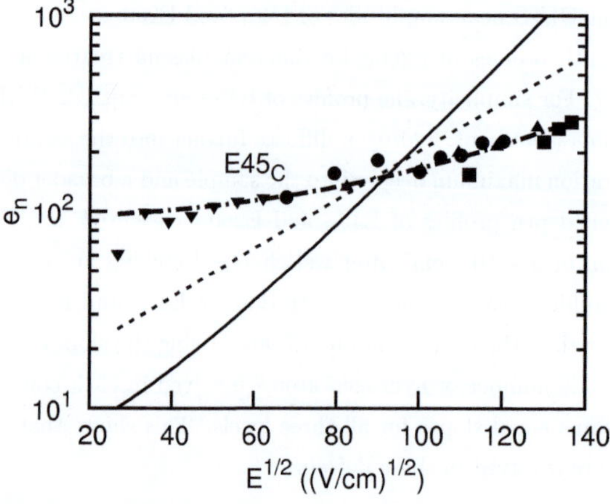

Figure 5.5: Electric field dependence of E45$_C$ taken at 45 K in different samples. Each data set is presented by a different symbol. Also shown are fits with the Hartke model (solid line), the 1D-Poole-Frenkel model (dotted line), and with the square well potential (dash-dotted line).

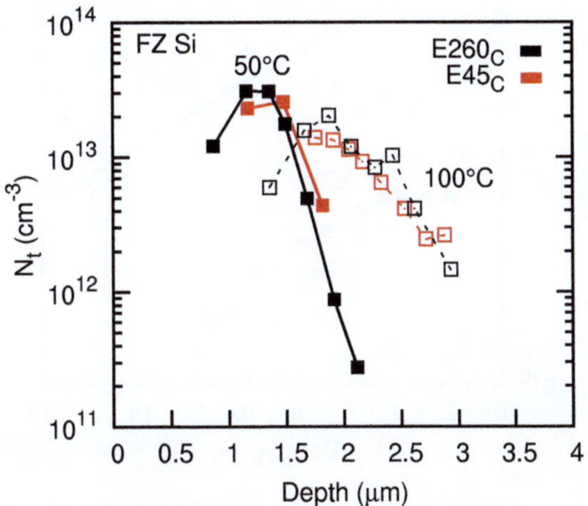

Figure 5.6: Concentration depth profiles of $E260_C$ and $E45_C$ in FZ silicon after hydrogen plasma at 50 °C (closed symbols) and 100 °C (open symbols).

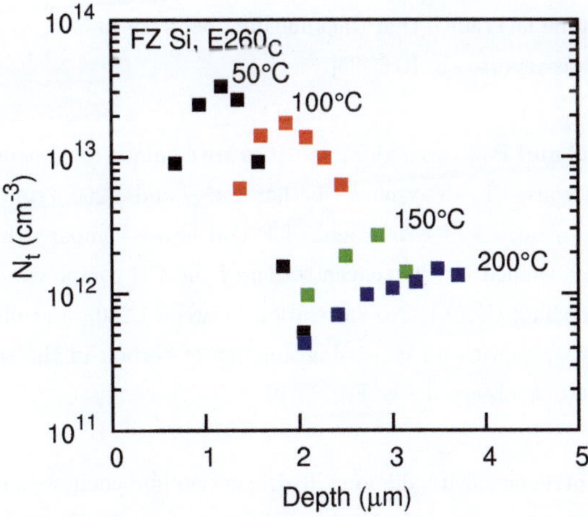

Figure 5.7: Concentration depth profiles of $E260_C$ in FZ silicon after hydrogen plasma at different temperatures.

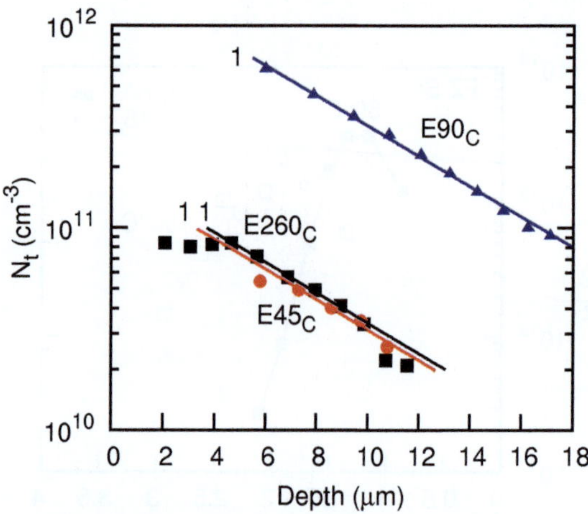

Figure 5.8: Depth profiles of E260$_C$, E90$_C$, and E45$_C$ in FZ silicon after RT wet chemical etching. The slopes of the concentrations tails are also presented.

Annealing The annealing behavior of E45$_C$ and E260$_C$ is shown in Fig. 5.9. Both peaks behave identical and disappear at 500 K. The other peaks observed directly after etching are less stable. E90$_C$ disappears already at 375 K, while E65$_C$ and E75$_C$ are stable up to 400 K. It should also be mentioned that the annealing behavior of E75$_C$ is strongly sensitive to illumination, as was reported in Ref. [35].

Correlation with C and P Two major impurities are common to all samples investigated: carbon and phosphorous. To determine whether E45$_C$ and E260$_C$ contain one of these impurities, the concentrations of E260$_C$ and of P and C are compared in Fig. 5.10. The carbon content is determined by the concentration of the CH complexes observed directly after wet chemical etching (E90$_C$). No correlation between E260$_C$ and phosphorous exists (see Fig. 5.10). However, with an increasing amount of carbon in the samples, a higher concentration of E260$_C$ is observed (see Fig. 5.10).

Isotope shift Experiments with different hydrogen isotope compositions in the plasma were performed. Figure 5.11 shows the Laplace DLTS spectra of E90$_C$ and E260$_C$ measured each in three samples of the same material, treated with either hydrogen, deuterium, or a hydrogen-deuterium mixture (50%-50%) and recorded under identical measurement conditions. For both defects, the peak position shifts depending on the hydrogen isotope used.

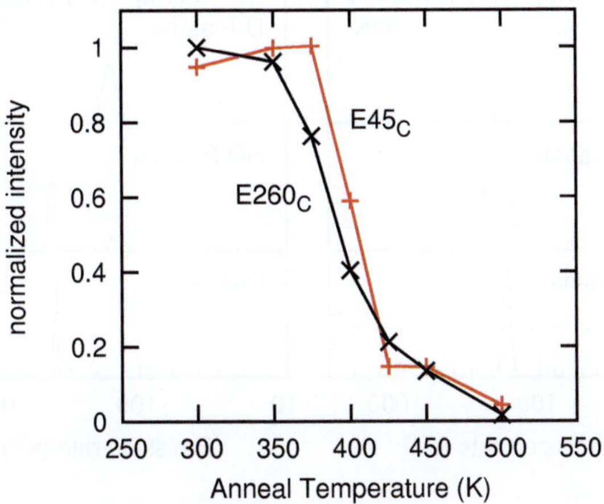

Figure 5.9: Annealing behavior of E45$_C$ and E260$_C$ taken from the DLTS peak amplitude from a sample treated with hydrogen plasma at 100 °C. The measurements were performed with $V_R = -2\,$V and $V_P = 0\,$V.

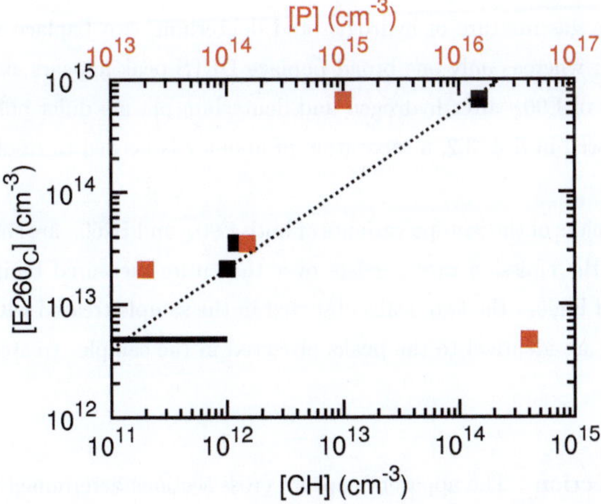

Figure 5.10: Dependence of the concentration of E260$_C$ on the concentration of phosphorous (red) and carbon (black). The carbon concentration is determined by the CH-concentration. For one sample, the CH-concentration was below the detection limit. This data point is shown as a line up to the upper limit of the CH-concentration. The dotted line is a guide to the eye.

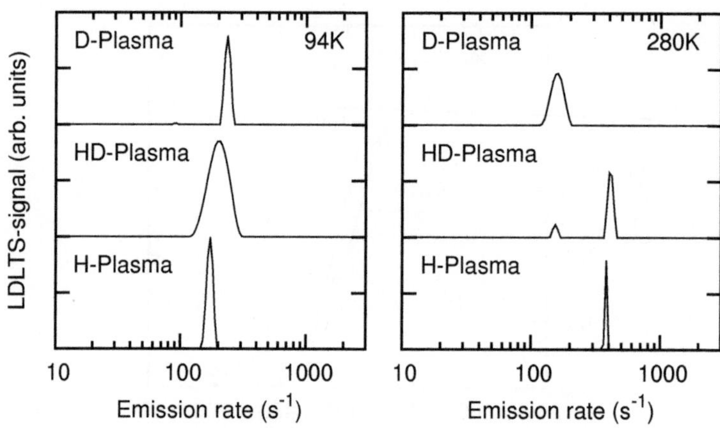

Figure 5.11: Laplace DLTS spectra recorded at 94 K and 280 K in FZ samples treated with hydrogen, hydrogen-deuterium (50%-50%), and deuterium plasma at 100 °C. The measurements were performed with $V_R = -4$ V and $V_P = -1$ V.

The use of deuterium increases the emission rate of $E90_C$, whereas it lowers the emission rate of $E260_C$. Using a gas mixture of hydrogen and deuterium, two Laplace DLTS peaks are observed at 280 K, whereas only one broad Laplace DLTS peak appears at 94 K. However, the emission rates of $E90_C$ after hydrogen and deuterium plasma differ only by a factor of 1.3. As was discussed in Sec. 3.2, a separation of about 2 is needed to resolve two peaks in Laplace DLTS.

The Arrhenius plots of the isotope variants of both $E90_C$ and $E260_C$ are shown in Fig. 5.12. The difference in the emission rate persists over the entire measured temperature region. Also, in the case of $E260_C$, the two peaks observed in the sample treated with the hydrogen-deuterium plasma are identical to the peaks observed in the samples treated with only one isotope.

Capture cross section   The apparent capture cross sections determined from the Arrhenius plot are found to be different for the hydrogen and deuterium related defects. This different behavior of the hydrogen and deuterium peaks is confirmed by direct capture measurements of $E260_C$, as is shown in Fig. 5.13. The directly measured capture cross sections exhibit a temperature dependence, which is presented in Fig. 5.14. This dependence yields a capture barrier of $E_\sigma = 30 \pm 5$ meV for both isotopes. Similar to the results from the

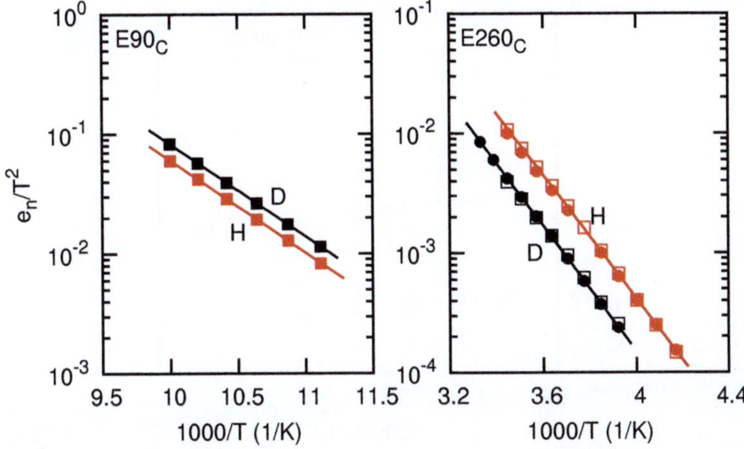

Figure 5.12: Arrhenius plots of the hydrogen and deuterium components of $E90_C$ and $E260_C$. Data points belonging to hydrogen are plotted in red, those belonging to deuterium in black. Closed symbols are measurements performed on samples treated with pure hydrogen or deuterium plasma. Open symbols are from a sample treated with a hydrogen-deuterium (50%-50%) plasma ($E260_C$ only).

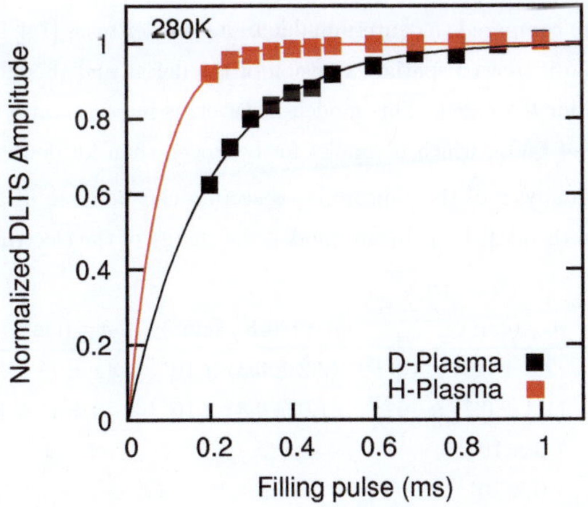

Figure 5.13: Direct capture measurement taken at 280 K on two samples of the same material, one treated with hydrogen plasma (red curve) and the other with deuterium plasma (black curve). The points are the measured data, the curves show the fitted capture rates.

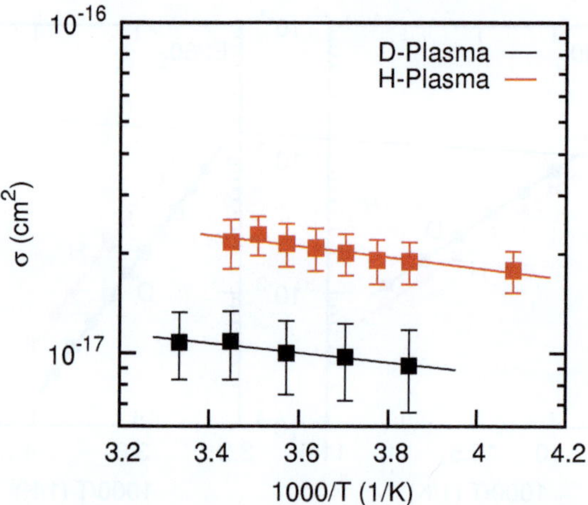

Figure 5.14: Determination of the capture barrier of E260$_C$ from the directly measured capture rates for deuterium (black) and hydrogen (red) plasma. For the obtained values see the text.

Arrhenius plot, the prefactor σ_∞ differs. An overview of the values for the capture cross section is given in Table 5.2.

A simple explanation of the different capture cross sections could be the larger vibrational amplitude of hydrogen compared to deuterium due to its smaller mass.[74] The larger amplitude translates into an increased spatial extension of the defect and therefore into a bigger capture cross section for hydrogen. This model, however, is inconsistent with the observed capture cross section of E90$_C$, which is smaller for hydrogen than for deuterium.

A more thorough analysis of the capture cross section can be done in the frame of the multiphonon emission theory.[54–57] In this model, the energy of the electron is transformed

isotope	σ_{na} (cm²)	$\sigma_n(280\ \mathrm{K})$ (cm²)	σ_∞ (cm²)
E260$_C$ H	$(2.8 \pm 0.9) \times 10^{-15}$	$(2.2 \pm 0.3) \times 10^{-17}$	$(8 \pm 1) \times 10^{-17}$
E260$_C$ D	$(1.4 \pm 0.7) \times 10^{-15}$	$(1.0 \pm 0.3) \times 10^{-17}$	$(3 \pm 1) \times 10^{-17}$
E90$_C$ H	1.3×10^{-15}	-	-
E90$_C$ D	1.6×10^{-15}	-	-

Table 5.2: Capture cross sections of E260$_C$ and E90$_C$ with hydrogen and deuterium. The apparent capture cross section σ_{na} is calculated from the Arrhenius plot. σ_n is measured directly at 280 K, and σ_∞ results from the barrier calculations.

Peak	E_{na}	kT	$\omega\ (kT)$	p	S	$\frac{\sigma_H}{\sigma_D}$ (theor.)	$\frac{\sigma_H}{\sigma_D}$ (exp.)
E260$_\mathrm{C}$	500 meV	25 meV	5	4	1	2.2	2-3
E90$_\mathrm{C}$	160 meV	8 meV	16	1.3	1	0.83	0.81

Table 5.3: Parameters for the capture cross section calculations for E260$_\mathrm{C}$ and E90$_\mathrm{C}$ for hydrogen (approximated). The C-H wag mode frequency is used ($\omega \approx 125$ meV). Both the calculated and the experimental capture cross section ratios are given.

upon capture into vibrational energy of the defect and dissipated into the crystal by the emission of phonons. Following the analysis of BOURGOIN AND LANNOO [57], the capture cross section σ can be described by

$$\sigma = \sigma_0 (\overline{n}+1)^p \frac{S^p}{p!} \exp\left[-2S\left(\overline{n}+\frac{1}{2}\right)\right] \tag{5.1}$$

with

$$\overline{n} = \left[\exp\left(\frac{\hbar\omega}{kT}\right)-1\right]^{-1} \tag{5.2}$$

and

$$\sigma_0 \propto \frac{1}{\omega} \tag{5.3}$$

where ω is the vibrational frequency of the defect, $\overline{n}$ is the average quantum number of the ground state (after capture), S is the HUANG-RHYS factor measuring the relative strength of the electron-phonon interaction, and p describes the energy difference before and after capture in units of $\hbar\omega$.

Using the measured activation energies and the C-H wag mode frequency [75], which results in $\omega \approx 125$ meV, the ratio $\frac{\sigma_H}{\sigma_D}$ is calculated from Eq. 5.1 for both defects. The ratio $\frac{\omega_h}{\omega_D}$ is taken for simplicity as $\sqrt{2}$. The corresponding parameters (for H) and both the calculated and experimental capture cross section ratios are given in Table 5.3. The calculated values match the experimental ones. An increase in the capture cross section for deuterium, as is observed for E90$_\mathrm{C}$, can be explained with this model.

The vibrational energy of the defect after capturing an electron has to be dissipated by the emission of phonons. Therefore, the lifetime of the vibrational mode after electron capture might have an additional influence on the capture cross section. It has been recently shown by both theory and vibrational spectroscopy that this lifetime can strongly depend on the isotope composition of the defect.[75–77] In the presence of a near-resonant mode, the isotope shift can lead to a mismatch of the phonon energies, requiring higher order phonon processes. Lifetime differences of an order of magnitude were reported.[75–77]

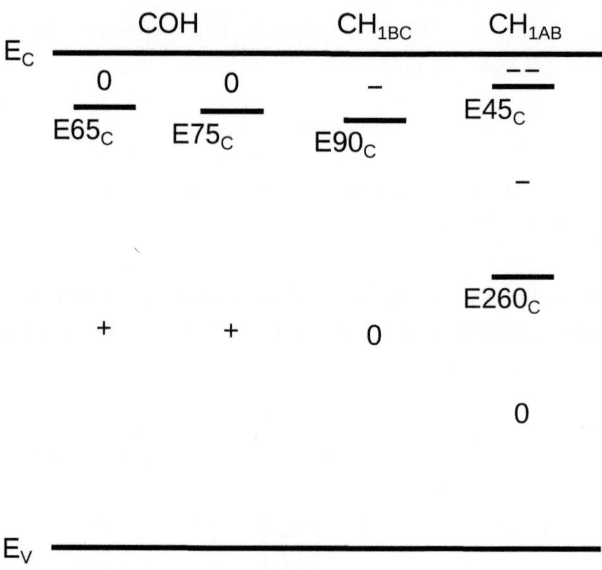

Figure 5.15: Overview of the electric levels of the hydrogen complexes in metal-free n-type silicon with their charge states. The picture is to scale.

5.3 Discussion

Figure 5.15 shows an overview of the levels of the hydrogen complexes observed in this chapter with their charge states and the assignment to the corresponding defects. The minor peaks $E90_C$, $E75_C$, and $E65_C$ observed after wet-chemical etching can be assigned to defects known from the literature. The origin of $E45_C$ and $E260_C$ is discussed below.

$E90_C$, which is present in FZ silicon samples, can be attributed to the carbon-hydrogen complex CH_{1BC}.[36] Activation energy and annealing behavior match those previously reported. As mentioned in the introduction of this chapter, STÜBNER et al.[73] showed that it is not the same configuration as the CH complex first reported by ENDRÖS et al.[37]

The peaks $E65_C$ and $E75_C$ observed in CZ silicon have properties which are very similar to the defects E1 and E2 reported by YONETA et al. and others.[21, 35] They were tentatively attributed to COH complexes.[35] The involvement of oxygen in these traps is highlighted by the fact that they only appear in CZ grown silicon, which contains more oxygen than FZ silicon.

$E45_C$ and $E260_C$ can be attributed to two charge states of the same defect. They appear together in both FZ and CZ material and have matching concentration profiles regardless of the sample preparation. Furthermore, their annealing behavior is identical ans is similar to the one reported before [39]. From the dependence of their emission rates on an applied electric field, $E45_C$ and $E260_C$ are assigned to a double acceptor and a single acceptor level, respectively. This is in agreement with JOHNSON *et al.*, who reported no POOLE-FRENKEL effect for either $E45_C$ or $E260_C$.[39] Further support comes from the higher apparent capture cross section of $E260_C$ compared to $E45_C$ (see Table 5.1).

The concentration depth profiles after wet chemical etching show the presence of only one hydrogen atom in the defect. Furthermore, the dependence of the concentration of $E260_C$ on the carbon content in the samples suggests the involvement of carbon in the defect structure. Therefore, this defect is attributed to a configuration of the CH complex.

Comparing the experimental data with theory, the most likely candidate for the observed defect is hydrogen at the anti-bonding position bound to carbon, CH_{1AB}.[36] Its formation energy is only slightly higher (0.2 eV) than that of CH_{1BC}. Also, it was calculated to exhibit an acceptor level at E_C - 0.63 eV [36], which is in good agreement to the position of $E260_C$ at E_C - 0.50 eV.

Furthermore, for the DLTS peaks $E260_C$ and $E90_C$ an influence of the hydrogen isotope on the capture cross section is observed. The observed effect can be described within the multiphonon emission theory by a change of the vibrational energy of the defect for the different masses. This implies that the captured electron is localized at the hydrogen atom. For a deeper understanding of the isotope effect on the capture cross section, including a possible influence of the vibrational lifetimes, a dedicated theoretical study would be helpful.

5.4 Summary

The investigation of hydrogenated metal-free silicon revealed the presence of several carbon and hydrogen related defects in silicon. The CH_{1BC} defect, observed as $E90_C$, is ubiquitous in etched n-type FZ silicon, whereas the COH defects $E65_C$ and $E75_C$ are common in etched n-type CZ silicon. They are not created by hydrogen plasma treatment at temperatures above 100 °C, but then $E45_C$ and $E260_C$ are observed in all samples. These peaks could be attributed to another CH-configuration, most likely CH_{1AB}.

Chapter 6

Titanium in silicon

This chapter focuses on the electrical properties of titanium in silicon. A controversy between the literature and the presented data on the charge states of the titanium peaks will be discussed. Also, several complexes of titanium with hydrogen are identified.

6.1   Introduction

Titanium is a very stable contaminant in silicon with a diffusion coefficient of about 3×10^{-9} cm^{2}s^{-1} at 1100 °C.[78] A reduction of solar cell performance in the presence of titanium is reported.[1, 79] Titanium exhibits three levels in the band gap, two in the upper half at E_C - 0.08 eV (E40$_{Ti}$) and E_C - 0.27 eV (E150$_{Ti}$) and one in the lower half at E_V + 0.29 eV (H180$_{Ti}$).[80–83] They were assigned to the single acceptor, single donor, and double donor of interstitial titanium, respectively.

The interaction of hydrogen with titanium was only studied in a few works. SINGH *et al.* presented data on proton implanted silicon.[34] In this study, the damaged surface layer after implantation was etched away and the sample was annealed at 300 °C. Neither TiH complexes nor a passivation of Ti could be observed. In contrast, JOST *et al.* showed the presence of two electrically active TiH complexes at E_C - 0.31 eV (E170$_{Ti}$) and E_C - 0.57 eV (E260$_{Ti}$).[31] Also, a passivation of Ti by hydrogen was observed and attributed to TiH$_4$.[31] Similar results were obtained by LEONARD *et al.*[84]

Theoretical studies on the interaction of hydrogen with titanium were recently published by BACKLUND AND ESTREICHER.[27] In their first principle studies, two configurations of Ti$_i$H were shown to exist. One complex has trigonal symmetry with Ti at a T site and H bound to Ti just beyond the nearest hexagonal interstitial site. In the second configuration (Ti$_i$H$_{BC}$) no bond between Ti and H exists. A Si atom near Ti moves away from the substitutional site along a trigonal axis, Ti$_i$ moves off the T site and overlaps with it, while

H ties up the Si dangling bond. Both configurations should introduce donor levels into the band gap of Si. The Ti_iH complex has a double donor level located at around $E_V + 0.13$ eV and a single donor level located at E_C - 0.38 eV. The single donor level of Ti_iH_{BC} lies at about E_C - 0.62 eV whereas its double donor level was shown to be deeper in the band gap at around $E_V + 0.42$ eV. The capture of a second hydrogen atom by a Ti atom was predicted to be energetically unfavorable since the formation of a H_2 molecule needs significantly less energy than that of the neutral charge state of Ti_iH_2.[27]

Besides Ti_iH-related defects, BACKLUND AND ESTREICHER demonstrated that also substitutional Ti (Ti_s) should form a complex with a single hydrogen atom.[27] This complex (Ti_sH) is supposed to introduce three levels into the band gap of Si: a single donor level at $E_V + 0.07$ eV, a single acceptor at E_C - 0.21 eV and a double acceptor at E_C - 0.10 eV. None of these levels was previously observed by means of the DLTS technique. The trapping of the second H atom by Ti_sH was shown to result in the movement of a Ti atom away from the substitutional site towards the T site. In this case, another complex defect (Ti_i+VH_2) has to be formed. This defect was predicted to introduce a donor and an acceptor level at $E_V + 0.05$ eV and E_C - 0.20 eV, respectively.[27]

In addition, the titanium-hydrogen interaction has recently been recalculated by SANTOS et al.[28] In contrast to the results of Ref. [27], they only found one stable TiH defect. The Ti_iH_{BC} defect was calculated to be metastable. For the Ti_iH configuration, SANTOS et al. reported three levels in the band gap: a single acceptor at E_C - 0.13 eV, a single donor at E_C - 0.48 eV, and a double donor at $E_V + 0.30$ eV. Moreover, complexes of titanium with two and three hydrogen atoms were reported to be stable in intrinsic or n-type silicon. For TiH_2, a single donor level at E_C - 0.50 eV and a double donor at $E_V + 0.36$ eV were calculated, whereas TiH_3 was shown to introduce a single donor level at E_C - 0.39 eV.[28] No stable substitutional titanium was found by their calculations.[28]

According to the results of Refs. [27, 28] one electrically inactive Ti-H complex, Ti_iH_4, should also be stable in Si. The presence of Ti_iH_4 was found to be consistent with the results previously reported.[31]

6.2 Results

DLTS Figure 6.1 presents the DLTS spectra recorded in n-type silicon samples which were doped with titanium during growth. The samples were either wet-chemically etched (blue lines) or treated with a DC hydrogen plasma (magenta lines). In the region close to the surface (Fig. 6.1 (a)), three DLTS peaks $E40_{Ti}$, $E150_{Ti}$, and $E260$ are detected in the wet-chemically etched samples. $E40_{Ti}$ and $E150_{Ti}$ are quite broad and consist of several peaks.

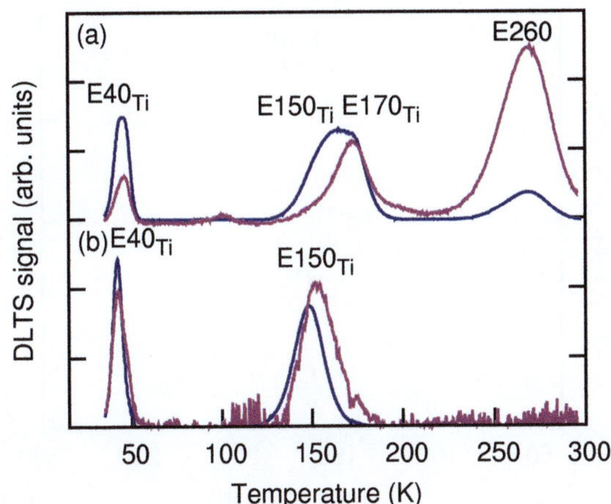

Figure 6.1: DLTS spectra of n-type silicon doped with titanium during growth after wet-chemical etching (blue line) and after hydrogen plasma treatment (magenta line). Shown are both spectra measured (a) close to the surface ($V_R = -2$ V and $V_P = 0$ V) and (b) in the bulk of the sample ($V_R = -8$ V and $V_P = -4$ V). The spectra were measured with a filling pulse width of 1 ms and a rate window of 47 s^{-1}.

In the plasma treated samples, the peak E260 dominates the spectrum. Instead of E150$_{Ti}$, a new sharp DLTS peak E170$_{Ti}$ is observed. Measuring deeper in the bulk (Fig. 6.1 (b)), only two sharp peaks E40$_{Ti}$ and E150$_{Ti}$ remain in both types of samples.

DLTS spectra of p-type silicon show only one peak H180$_{Ti}$ (see Fig. 6.2). No additional lines appear after hydrogen introduction by either etching or hydrogen plasma treatment.

MCTS Via MCTS, the presence of the other two Ti peaks, E40$_{Ti}$ and E150$_{Ti}$, in p-type silicon was verified. The measured MCTS spectrum is presented in Fig. 6.3 together with a DLTS spectrum of n-type silicon. Both peaks E40$_{Ti}$ and E150$_{Ti}$ are also present in the p-type material.

Laplace DLTS Laplace DLTS measurements on the broad peaks in n-type Si close to the surface reveal that they consist in fact of several peaks. Figure 6.4 shows the Laplace DLTS results measured at 45 K after hydrogen plasma treatment. Two peaks are observed, labeled E40$_{Ti}$ and E40'$_{Ti}$. The same peaks are also observed in wet-chemically etched samples. Their intensity ratio varies, with a higher intensity of E40'$_{Ti}$ after more hydrogen is introduced into the sample.

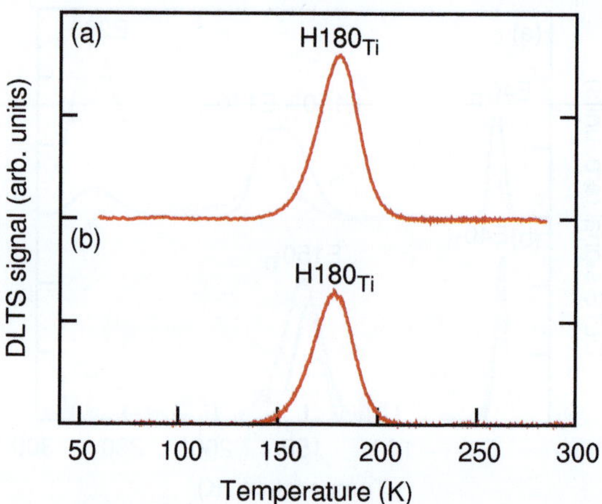

Figure 6.2: DLTS spectra of p-type silicon doped with titanium during growth after wet-chemical etching. Shown are spectra measured (a) close to the surface ($V_R = -1$ V and $V_P = 0$ V) and (b) in the bulk of the sample ($V_R = -6$ V and $V_P = -2$ V). The spectra were measured with a filling pulse width of 1 ms and a rate window of 47s $^{-1}$.

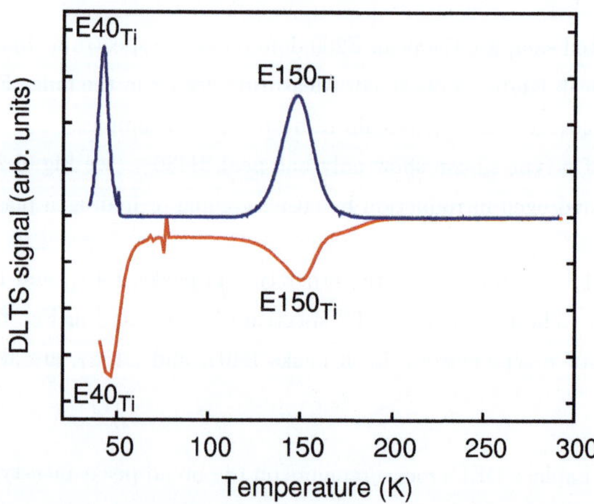

Figure 6.3: Comparison of an MCTS spectrum of p-type silicon (red line) with a DLTS spectrum of n-type silicon (blue line), both containing titanium. The measurement parameters for MCTS were $V_R = -6$ V, a laser pulse width of 10 ms, and a rate window of 80 s^{-1}. The DLTS spectrum was taken with $V_R = -8$ V, $V_P = -4$ V), a filling pulse width of 1 ms, and a rate window of 47 s^{-1}.

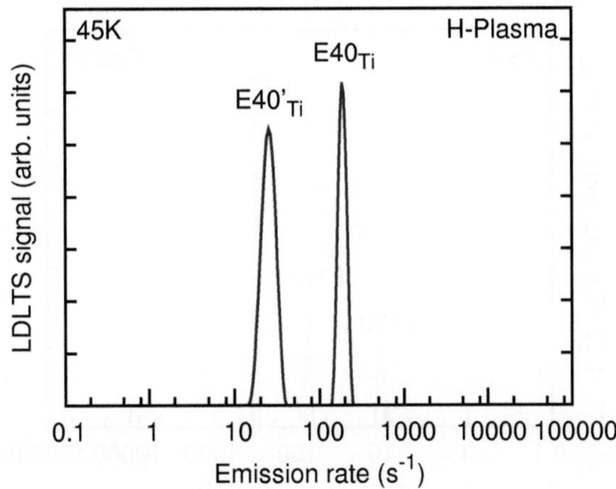

Figure 6.4: Laplace DLTS spectra recorded at 45 K in Ti-doped n-type silicon after hydrogen plasma treatment.

The Laplace DLTS data measured at 165 K in samples treated with hydrogen plasma are presented in Fig. 6.5. One can clearly see the presence of three sharp Laplace DLTS peaks, labeled E150$_{Ti}$, E170$_{Ti}$, and E170'$_{Ti}$. The intensities of E170$_{Ti}$ and E170'$_{Ti}$ are higher in respect to E150$_{Ti}$ in samples treated with hydrogen plasma than in etched samples.

Figure 6.6 shows the Laplace DLTS spectra recorded at 280 K in samples after wet-chemical etching and after hydrogen plasma treatment. The etched sample shows only one Laplace DLTS peak. In the plasma treated samples, however, E260 consists of two Laplace DLTS peaks, labeled E260$_{Ti}$ and E260$_{C}$. E260$_{C}$ is identical to the CH-peak discussed in Chap. 5.

Arrhenius plot The activation energies and apparent capture cross sections were determined from the Arrhenius plots for all defects presented in Fig. 6.7. The values are listed in Table 6.1 together with the defect assignment, which is argued in the discussion.

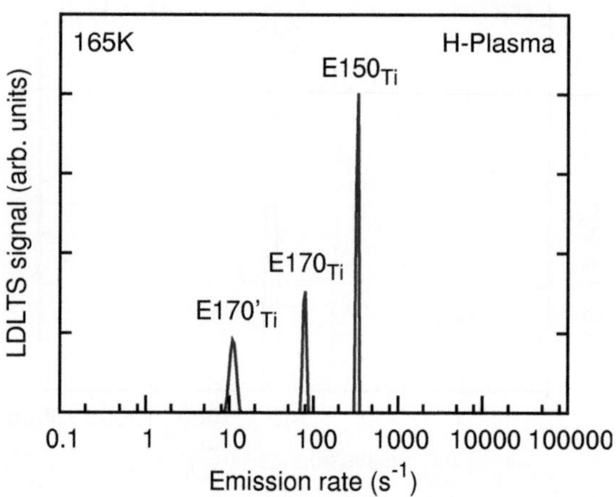

Figure 6.5: Laplace DLTS spectra recorded at 165 K in Ti-doped n-type silicon after hydrogen plasma treatment.

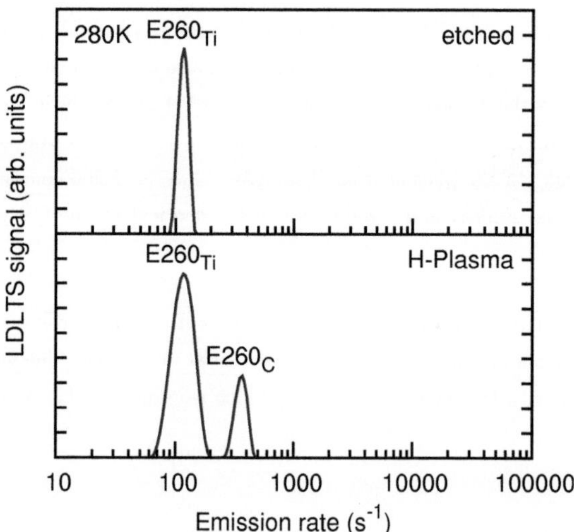

Figure 6.6: Laplace DLTS spectra recorded at 280 K in Ti-doped n-type silicon after wet-chemical etching (top) and after hydrogen plasma treatment (bottom).

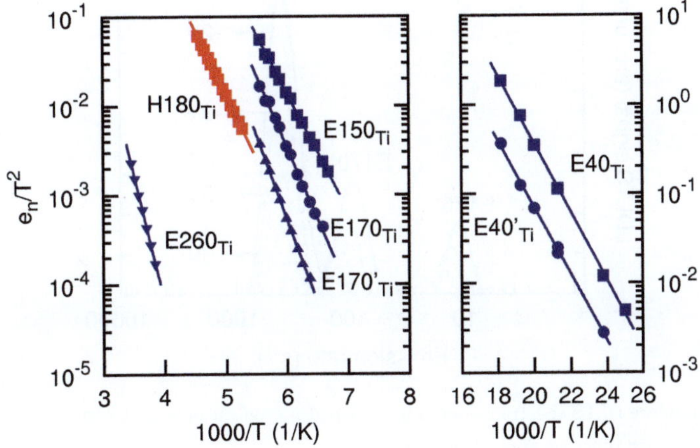

Figure 6.7: Arrhenius plots of all Ti-related peaks observed. The activation energies and apparent capture cross sections are listed in Table 6.1.

label	E_{na} (eV)	σ_{na} (cm^2)	assignment
E260$_{Ti}$	-0.55	4×10^{-15}	TiH$_3$ (?/?)
E170$_{Ti}$	-0.34	4×10^{-14}	TiH$_{BC}$ (0/+)
E170'$_{Ti}$	-0.37	4×10^{-14}	TiH$_2$ (0/+)
E150$_{Ti}$	-0.27	5×10^{-16}	Ti (-/0)
E40$_{Ti}$	-0.08	6×10^{-15}	Ti (0/+)
E40'$_{Ti}$	-0.07	8×10^{-16}	TiH (0/+)
H180$_{Ti}$	$+0.29$	9×10^{-17}	Ti (+/++)

Table 6.1: Activation energies, apparent capture cross sections, and assignment for all Ti-related defects seen in n- and p-type Si.

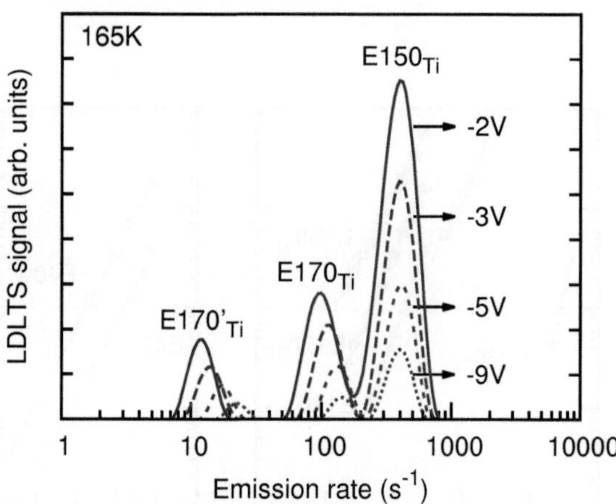

Figure 6.8: Laplace DLTS spectra recorded in Ti-doped n-type silicon after hydrogen plasma treatment at 165 K with different applied reverse bias.

Electric field dependence To determine the charge states of the observed DLTS peaks, the influence of the electric field on the emission rate of the defects was investigated. Figure 6.8 shows the Laplace DLTS spectra taken at 165 K with different applied reverse bias. The emission rate of E150$_{Ti}$ does not shift with an increase in the applied electric field, whereas both E170$_{Ti}$ and E170'$_{Ti}$ shift to higher emission rates.

To check whether this shift of E170$_{Ti}$ and E170'$_{Ti}$ corresponds to the POOLE-FRENKEL effect, the logarithm of the emission rate is plotted versus the square root of the applied electric field in Fig. 6.9. As one can see, the data shows excellent agreement with the theoretically predicted behavior calculated from both the HARTKE model (Eq. 2.24) and the 1D-POOLE-FRENKEL model (Eq. 2.23). Therefore, both defects can be assigned to a donor state in the upper half of the band gap.

The dependence of E40$_{Ti}$ on the electric field is presented in Fig. 6.10 together with the theoretical curves from the 1D-POOLE-FRENKEL model (dotted line) and from the HARTKE model (solid line). The data is in good agreement with the behavior expected in the 1D-POOLE-FRENKEL model. E40$_{Ti}$ is therefore assigned to a donor level.

Figure 6.11 shows the electric field dependence of E40'$_{Ti}$. For an electric field lower than 1×10^4 V/cm, the data shows a linear dependence on the square root of the electric field. For higher fields, the data points deviate from this line. In this field region, the data points show a linear dependence on the square of the electric field (see inset in Fig. 6.11). This behavior

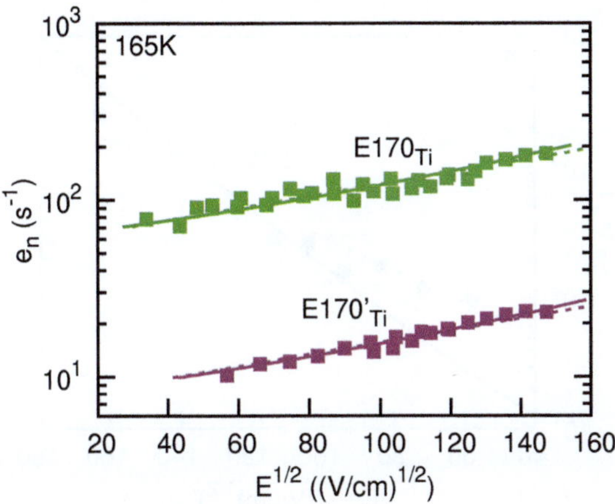

Figure 6.9: Emission rates of E170 and E170' versus the square root of the electric field measured at 165 K. The data are fitted with the HARTKE model (solid line) and the 1D-POOLE-FRENKEL model (dotted line).

can be explained by the onset of tunneling at sufficiently high electric fields, as was reported in Ref. [61] One should also note that the slope of the dependence on the square root of the field is smaller by a factor of 3-4 than expected from the POOLE-FRENKEL theory. However, similarly reduced slopes have been reported before for Coulombic attractive centers.[61] E40'$_{Ti}$ can therefore also be assigned to a donor level.

Depth profiles Figure 6.12 shows the depth profiles of the defects observed on n-type silicon samples after wet chemical etching. E40$_{Ti}$ and E150$_{Ti}$ have constant and equal concentrations in the bulk of the sample. Both defect concentrations decrease towards the surface. The concentration of E40$_{Ti}$, however, decreases stronger than that of E150. The concentrations of E40'$_{Ti}$, E170$_{Ti}$, E170'$_{Ti}$, and E260$_{Ti}$ are highest near the surface and drop towards the bulk. To determine the microscopic structure of the defects, the method outlined in Sec. 2.4 is applied. The slopes have a ratio of about 1:1:2:3 for the defects E40'$_{Ti}$,E170$_{Ti}$, E170'$_{Ti}$, and E260$_{Ti}$. This indicates the presence of 1, 2, and 3 hydrogen atoms in their respective structure.

To control the origin of the hydrogen related peaks, the concentrations of E40'$_{Ti}$, E170$_{Ti}$, E170'$_{Ti}$, and E260$_{Ti}$ were summed up with either E40$_{Ti}$ or with E150$_{Ti}$. The results are presented in Fig. 6.13. The sum with E40$_{Ti}$, shown as the black curve, is constant over the

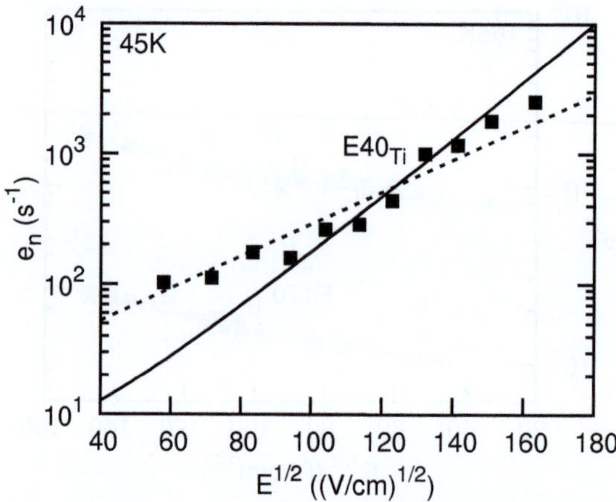

Figure 6.10: Emission rate of E40$_{Ti}$ versus the square root of the electric field measured at 45 K. The data are fitted with the HARTKE model (solid line) and the 1D-POOLE-FRENKEL model (dotted line).

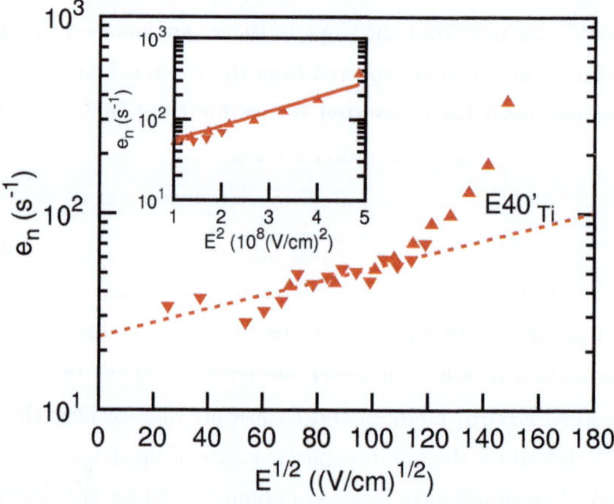

Figure 6.11: Emission rate of E40'$_{Ti}$ versus the square root of the electric field measured at 45 K. Two datasets marked with different symbols are combined in this graph. The dotted line shows a linear fit of the experimental points at fields below 1×10^4 V/cm. The inset shows the dependence on E^2 for electric fields higher than 1×10^4 V/cm with a linear fit (solid line).

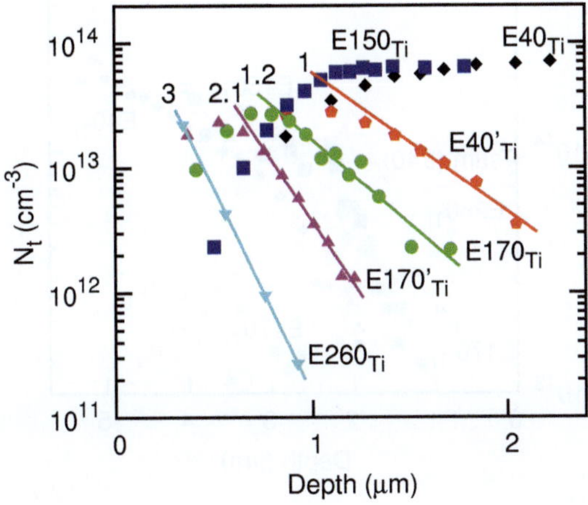

Figure 6.12: Depth profile of the defects observed in Ti-doped n-type samples after etching. The fits to the concentration tails and their slopes normalized to the slope of E40'$_{Ti}$ are also shown.

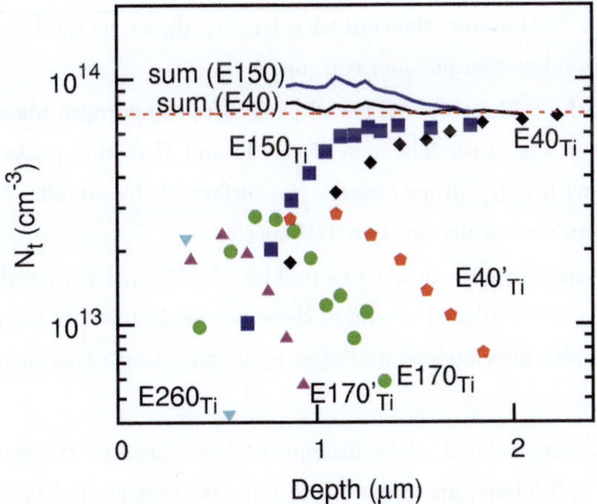

Figure 6.13: Depth profile of the defects observed in Ti-doped n-type Si after etching. The full-drawn lines depict the summed-up profiles of the H-related peaks E40'$_{Ti}$, E170$_{Ti}$, E170'$_{Ti}$, and E260$_{Ti}$ with either E40$_{Ti}$ (black line) or with E150$_{Ti}$ (blue line). The dotted red line is a guide to the eye at a constant value of 7.3×10^{13}cm^{-3}.

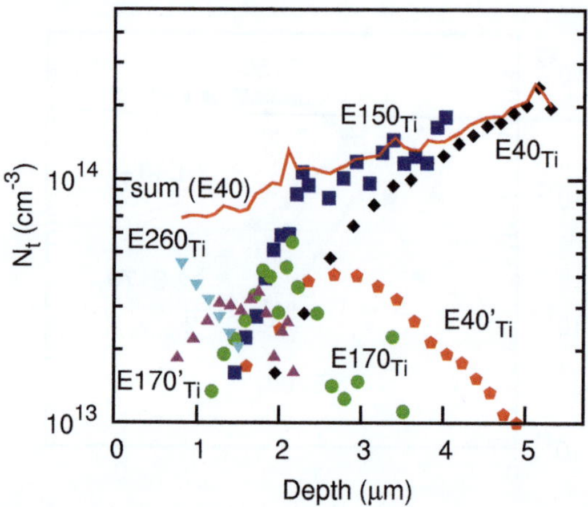

Figure 6.14: Depth profiles of the Ti-related defects observed in n-type samples after hydrogen plasma treatment. The sum of the TiH-related peaks E40'$_{Ti}$, E170$_{Ti}$, E170'$_{Ti}$, and E260$_{Ti}$ with E40$_{Ti}$ (red line) is also shown.

entire observed region. The dotted red line in Fig. 6.13 is a guide to the eye at a constant value of $7.3 \times 10^{13} \mathrm{cm}^{-3}$. However, the sum with E150$_{Ti}$, drawn as the blue line, exhibits a change in concentration between one and two µm.

Concentration depth profiles recorded for all peaks after a hydrogen plasma treatment at 100 °C are presented in Fig. 6.14. The sum of the Ti and H related peaks E40'$_{Ti}$, E170$_{Ti}$, E170'$_{Ti}$, and E260$_{Ti}$ with E40$_{Ti}$ drops towards the surface of the sample. This is consistent with the presence of an electrically inactive TiH-species.

Figure 6.15 shows the concentration depth profiles of E150$_{Ti}$, E40$_{Ti}$, and E40'$_{Ti}$ together with a sum over E40$_{Ti}$ and E40'$_{Ti}$. The sum of these two peaks matches the profile of E150$_{Ti}$ very well both in samples after etching and after hydrogen plasma treatment.

Annealing　All hydrogen related peaks disappear after annealing the samples at 250 °C. The Ti peaks E40$_{Ti}$ and E150$_{Ti}$ are stable up to temperatures around 900 °C. Figure 6.16 shows the depth profiles of E40 and E150 taken after a heat treatment at 300 °C to destroy the hydrogen complexes and after a 1 h heat treatment at 850 °C. As expected, annealing at 300 °C has no influence on the total Ti concentration in the sample, and the profile observed is flat. After annealing the sample at 850 °C, however, the Ti concentration in the sample is reduced. Moreover, it drops towards the surface, which might be due to an out-diffusion of

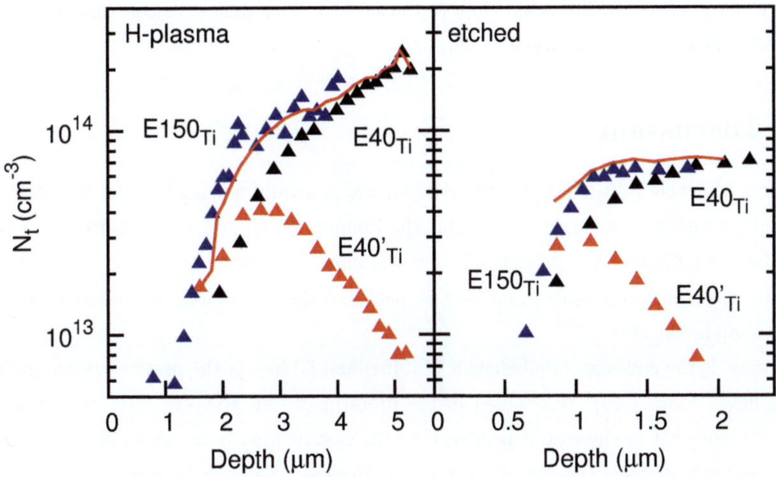

Figure 6.15: Concentration depth profiles of E40$_{Ti}$, E40'$_{Ti}$, and E150$_{Ti}$ in a hydrogen-plasma treated (left) and a wet-chemically etched (right) sample. The solid red line depicts the summed-up concentration of E40$_{Ti}$ and E40'$_{Ti}$.

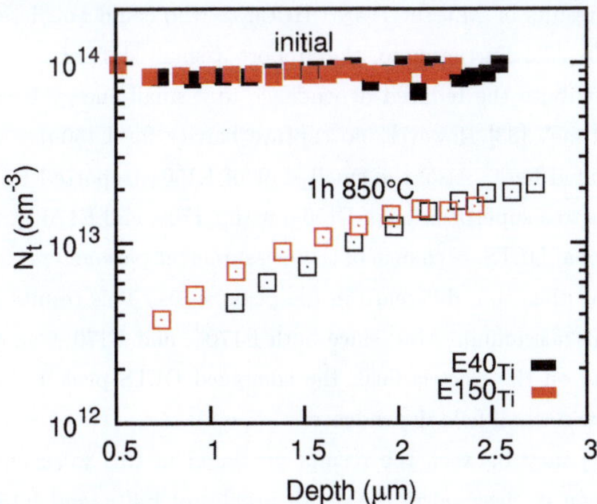

Figure 6.16: Depth profile of the Ti levels E40$_{Ti}$ and E150$_{Ti}$ observed in n-type samples after annealing at 300 °C (full symbols) and 850 °C (open symbols).

Ti in the sample. It is also interesting to note that the concentrations of $E40_{Ti}$ and $E150_{Ti}$ do not exactly match in the region close to the surface. This difference can consistently be observed after high temperature annealing.

6.3   Discussion

The three Ti levels $E40_{Ti}$, $E150_{Ti}$, and $H180_{Ti}$ are generally assigned in the literature to the single acceptor, the single donor, and the double donor level of interstitial Ti (Ti_i), respectively.[80–83] The data presented here reveal some inconsistencies in this model, namely the shift of the emission rates in an electric field and the differences in the concentration profiles close to the surface.

Foremost is the emission rate behavior of $E40_{Ti}$ and $E150_{Ti}$ in the presence of an applied electric field. Since $E40_{Ti}$ is assigned in the literature to an acceptor level in n-type Si, one would expect it to show no dependence on the electric field, whereas $E150_{Ti}$ as a donor level should exhibit the POOLE-FRENKEL effect. However, $E40_{Ti}$ does show an electric field dependence, and $E150_{Ti}$ does not, as can be seen in Figs. 6.8 and 6.10. The charge states inferred from this behavior would be the donor level for $E40_{Ti}$ and the acceptor level for $E150_{Ti}$. However, if both peaks belong to the same defect, this would result in a negative-U ordering, which is not likely for Ti in silicon [27].

It should be noted that the presence of an electric field dependence of $E40_{Ti}$ is in disagreement with the results of MATHIOT AND HOCINE, who could not detect a shift of the emission rate of $E40_{Ti}$.[83] Furthermore, they report a small shift of $E150_{Ti}$ with applied electric field, and attribute the reduced dependence to a small energy barrier for electron capture of about 20 meV.[83] However, no capture barrier for $E150_{Ti}$ could be found in the samples investigated here. Also, the small shift of $E150_{Ti}$ reported by MATHIOT AND HOCINE could be due to a superposition of $E150_{Ti}$ with $E170_{Ti}$ and $E170'_{Ti}$, which cannot be resolved by conventional DLTS. A change of the measurement parameters can lead to a shift of probed volume and thus to a difference in the peak ratios. This results in a shift of the combined DLTS peak maximum. Also, since both $E170_{Ti}$ and $E170'_{Ti}$ show a dependence of their emission rate on the electric field, the combined DLTS peak $E150_{Ti}$ + $E170_{Ti}$ + $E170'_{Ti}$ will also show a small field dependence.

The second discrepancy between the results presented in this work and the literature model for Ti_i in silicon is observed in the depth profiles of $E40_{Ti}$ and $E150_{Ti}$ close to the surface. Depth profiles after annealing the samples at temperatures around 800 °C to 950 °C show a slight difference in the concentrations of $E40_{Ti}$ and $E150_{Ti}$ (see Fig. 6.16). Also, after hydrogenation, $E150_{Ti}$ has higher concentrations close to the surface than $E40_{Ti}$ (see

Figs. 6.12, 6.13 and 6.14). This is consistent with the charge state assignment made above, as hydrogen is negatively charged in n-type silicon.[46, 85] Then, a reaction of hydrogen with the positively charged $E40_{Ti}$ is more likely to occur than with the neutral $E150_{Ti}$. It should be noted, though, that the difference in the concentration depth profile of $E40_{Ti}$ and $E150_{Ti}$ after hydrogenation could also be explained by the possible presence of a further DLTS peak which could not be resolved from $E150_{Ti}$. This idea is supported by the depth profiles presented in Fig. 6.15, where the sum of $E40_{Ti}$ and $E40'_{Ti}$ is similar to the profile of $E150_{Ti}$.

Furthermore, the concentration differences observed here only happen in regions of a few hundred nm. Measurements in the bulk, regardless of the treatment of the samples, show consistently equal concentrations of $E40_{Ti}$ and $E150_{Ti}$. This is in agreement with the results presented in the literature.[83, 86] If the two peaks would belong to two different defects, different bulk concentrations should be achievable.

The introduction of hydrogen into Ti-containing samples results in the appearance of several additional peaks, $E40'_{Ti}$, $E170_{Ti}$, $E170'_{Ti}$, $E260_{Ti}$, and $E260_{C}$.

$E260_{C}$ appears only after hydrogen plasma treatment. It is unrelated to titanium but is identical to the peak observed in metal-free samples. Its properties are discussed in detail in Chap. 5. The presence of $E260_{C}$ also explains why the intensity of E260 observed in the DLTS spectra after plasma hydrogenation is significantly larger than that of the Ti peaks.

The electrical properties of $E170_{Ti}$ and $E260_{Ti}$ were previously reported in Refs. [31, 84] where the defects were attributed to TiH-related defects. This assignment to TiH-related defects is also consistent with my observations. $E170_{Ti}$ and $E260_{Ti}$ appear after hydrogenation of the samples in the region close to the surface and no traces of the peak were observed deeper than 6 µm. In addition, their intensity is reversely correlated with that of interstitial Ti ($E40_{Ti}$) and increases with higher H content in the samples. Therefore, these peaks can be correlated with TiH-related defects. Similar conclusions can be drawn about the origin of $E40'_{Ti}$ and $E170'_{Ti}$. The relatively low thermal stability of the defects is in good agreement with this assignment.

As mentioned above, the slope of the concentration profiles was found to be identical for $E40'_{Ti}$ and $E170_{Ti}$ whereas it was twice as steep for $E170'_{Ti}$ and three times as steep for $E260_{Ti}$. Therefore, $E170'_{Ti}$ contains twice and $E260_{Ti}$ three times more hydrogen atoms than $E40'_{Ti}$ and $E170_{Ti}$. One should also notice that an equal number of hydrogen atoms does not necessarily signify that $E40'_{Ti}$ and $E170_{Ti}$ belong to different charge states of the same defect. In support of this idea, different depth profiles (see Fig. 6.12) confirm that $E40'_{Ti}$ and $E170_{Ti}$ originate from different defects in spite of the equal number of hydrogen atoms in their structure.

It seems very likely that E40'$_{\text{Ti}}$, E170$_{\text{Ti}}$, E170'$_{\text{Ti}}$, and E260$_{\text{Ti}}$ contain only one Ti atom. Evidence for this model is given by the constant sum over the concentrations of E40$_{\text{Ti}}$, E40'$_{\text{Ti}}$, E170$_{\text{Ti}}$, E170'$_{\text{Ti}}$, and E260$_{\text{Ti}}$ presented in Fig. 6.12. Another argument against the formation of an electrically active Ti$_y$H$_x$ complex with two or more Ti atoms is the Ti$_i$ concentration of only about 10^{14} cm^{-3} in these samples. At this low concentration, only a very small probability to form Ti$_y$H$_x$ complexes with $y > 1$ exists and no evidences of the presence of such complexes were previously found in the literature.

Therefore, E40'$_{\text{Ti}}$ and E170$_{\text{Ti}}$ are assigned to TiH defects while E170'$_{\text{Ti}}$ and E260$_{\text{Ti}}$ are attributed to a TiH$_2$ and a TiH$_3$ complex, respectively. The enhancement of the emission rates on the electric field (Figs. 6.9 and 6.10) allows the conclusion that the minor TiH-related peaks E40'$_{\text{Ti}}$, E170$_{\text{Ti}}$, and E170'$_{\text{Ti}}$ are single donors.

As mentioned in the introduction, two different configurations of the Ti$_i$H complex and one configuration of the Ti$_s$H complex were predicted to be stable in Si.[27] It seems therefore reasonable to attribute E40'$_{\text{Ti}}$ to the single donor level of Ti$_i$H and E170 to the single donor level of Ti$_i$H$_{\text{BC}}$. Ti$_s$H can be excluded, since the summed up concentrations match very well with E40$_{\text{Ti}}$ (Ti$_i$). Also, Ti$_s$ should exhibit acceptor levels in the upper half of the band gap,[27] whereas E40'$_{\text{Ti}}$ and E170$_{\text{Ti}}$ are donor levels. The assignment of E40'$_{\text{Ti}}$ and E170$_{\text{Ti}}$ to two different charge states of only one TiH configuration, as proposed by SANTOS $et\ al.$[28], is inconsistent with both the different depth profiles of these defects and their identical charge state.

Furthermore, E170'$_{\text{Ti}}$ is assigned to the single donor state of the TiH$_2$ defect. The presence of the TiH$_2$ defect is inconsistent with the results of ab-initio calculations where the formation of TiH$_2$ was predicted to be energetically less favorable than that of H$_2$ molecules.[27] The interaction between Ti and H in Ref. [27] was assumed to be in the neutral charge state for both species. However, this condition is not realized in the present study. Using capacitance studies, HERRING $et\ al.$ [46] and K. BONDE NIELSEN $et\ al.$ [85] demonstrated that isolated hydrogen should be negatively charged at and slightly above room temperature in n-type Si. The stability of negatively charged H in n-type Si was also confirmed by ab-initio calculations reported in Ref. [87].

The negative charge state of hydrogen was taken into account in the calculations of SANTOS $et\ al.$ which showed the presence of stable TiH$_2$ and TiH$_3$ complexes in n-type silicon.[28] Their calculated level positions for TiH$_2$ (E_C - 0.50 eV) and TiH$_3$ (E_C - 0.39 eV) are in rough agreement with the experimental results (E_C - 0.38 eV and E_C - 0.50 eV).

The reduction of the total electrically active defect concentration towards the surface after hydrogen plasma treatment (see Fig. 6.14) suggests the presence of an electrically inactive

TiH-complex. In accordance with the literature, [27, 28, 31] this passive complex might be attributed to TiH_4.

Finally, another titanium and hydrogen related defect might exist which cannot be resolved from $E150_{Ti}$. Indications that this might be the case have been presented in Fig. 6.15. From the sums presented there, one can assume that if such a hidden defect exists, it would have the same concentration depth profile as $E40'_{Ti}$. It would therefore be a second charge state of Ti_iH.

No additional TiH_x-related levels were found in p-type Si. These observations are consistent with those previously reported in Ref. [31]. These findings can be correlated with the identical positive charge state of H (see Refs. [46, 85]) and Ti which lead to Coulombic repulsion between these species. As a result, no passivation of $H180_{Ti}$ occurs in samples subjected to hydrogenation. This is also in agreement with the theoretical predictions by SANTOS et $al.$, who showed a reduced stability of TiH complexes in p-type silicon.[28]

At last, I would like to comment on the results of Ref. [34]. As mentioned above, SINGH et $al.$ reported that TiH-related complexes had not been observed after proton implantation and the following annealing at 570 K in n-type Si.[34] In contrast, four TiH-related DLTS peaks were detected in Figs. 6.1 to 6.6 in our samples. The observed discrepancy can be explained by the high annealing temperature of the samples immediately after proton implantation as reported in Ref. [34]. As mentioned above, the TiH-related defects ($E40'_{Ti}$, $E170_{Ti}$, $E170'_{Ti}$, and $E260_{Ti}$) anneal out after the heat treatment at 520 K. Therefore, the absence of passivation of isolated Ti in Ref. [34] can be directly correlated with the complete dissociation of the TiH_x complexes after the annealing at 570 K.

An overview of all the titanium-related levels determined in this work is given in Fig. 6.17. The dashed lines indicate the shift of the level energies by hydrogenation.

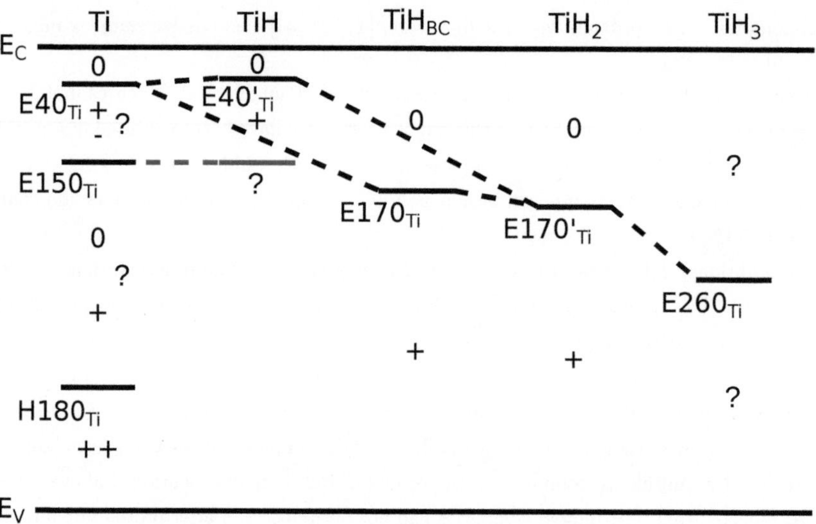

Figure 6.17: Overview of the electric levels of titanium and the TiH complexes in the band gap of silicon with their charge states. The picture is to scale.

6.4 Summary

The data presented in this chapter reveal inconsistencies of the literature assignment of the Ti levels $E40_{Ti}$, $E150_{Ti}$, and $H180_{Ti}$ to the single acceptor, the single donor, and the double donor level of interstitial Ti (Ti_i). The observed behavior of the emission rate in an applied electric field for $E40_{Ti}$ and $E150_{Ti}$ is in contrast to these charge states. Also, concentration differences near the surface can be observed for $E40_{Ti}$ and $E150_{Ti}$ after hydrogenation or annealing.

After an introduction of hydrogen into the samples, four electrically active TiH-complexes $E40'_{Ti}$, $E170_{Ti}$, $E170'_{Ti}$, and $E260_{Ti}$ were detected in n-type silicon. $E40'_{Ti}$ and $E170_{Ti}$ can be assigned to two different configurations of Ti_i with one hydrogen atom, as proposed by theory.[27] They are attributed to Ti_iH (E40') and Ti_iH_{BC} ($E170_{Ti}$). Possibly, a second level of Ti_iH exists which cannot clearly be resolved from $E150_{Ti}$.

The other defects are assigned to TiH_2 ($E170'_{Ti}$) and TiH_3 ($E260_{Ti}$). Additionally, in accordance with previous results [31] and theory [27, 28], a passive complex, probably TiH_4, was concluded to exist. No TiH-related defects were observed in p-type Si.

Chapter 7

Cobalt in silicon

This chapter focuses on cobalt in silicon and its interaction with hydrogen. The results of previous studies on the electrical properties of cobalt are controversial. With the results presented here, the assignment of the cobalt-related peaks is reconsidered.

7.1 Introduction

Cobalt was shown to introduce a deep acceptor in silicon at about E_C - 0.4 eV (E200$_{Co}$) [29, 88–90] whereas different donor levels of Co were reported at $E_V + 0.21$ eV [89, 90] and at $E_V + 0.4$ eV (H240$_{Co}$) [30, 88]. KITAGAWA tentatively assigned the acceptor level at E_C - 0.4 eV and the donor level at $E_V + 0.21$ eV to substitutional cobalt.[12] In contrast, SCHEIBE AND SCHRÖTER showed that substitutional cobalt does not introduce electrically active levels in the band gap from $E_V + 0.1$ eV to $E_V + 0.5$ eV.[91] The authors assigned the levels in the lower half of the band gap to defect complexes formed during the quenching of the sample.

Only about 0.1% of the total dissolved cobalt in silicon was found to be electrically active.[11, 12] Most of these electrically active species were assigned to substitutional cobalt.[12, 92] The annealing temperature of substitutional cobalt varies in different studies.[12, 92] Cobalt was shown to anneal around 400 °C by BERGHOLZ et $al.$ [92] whereas a higher temperature of about 600 °C was reported by KITAGAWA [12].

ZHANG et $al.$ have shown that at least in the neutral state, the formation energies for interstitial and substitutional cobalt are very close, with a preference for the interstitial site.[93] Using spin-unrestricted self-consistent electronic and magnetic GREEN-function calculations, BEELER et $al.$ also suggested that both interstitial and substitutional Co species should be present in Si.[94] In agreement with the results of the theoretical calculations, LEMKE AND IRMSCHER assigned two defect levels with the activation energies of $E_V + 0.31$ eV and

E_C - 0.4 eV to the donor and acceptor states of interstitial Co_i in addition to the substitutional Co_s levels at E_V + 0.4 eV and E_C - 0.4 eV.[95] However, direct evidence for interstitial Co_i is missing. The positions of the single acceptor and single donor levels of Co_s found by LEMKE AND IRMSCHER are in good agreement with those reported in Refs. [29, 30, 88].

In contrast to Ref. [95], other studies suggested interstitial cobalt to be unstable at room temperature.[92] Using combined Mössbauer spectroscopy and transmission electron microscopy cobalt was also shown to precipitate into $CoSi_2$ particels during quenching.[96]

Interstitial cobalt reacts with shallow acceptors forming cobalt-acceptor pairs. Boron, aluminum, and gallium pairs were reported by BERGHOLZ AND SCHRÖTER.[97] The presence of the CoB pair was also confirmed in p-type Si by SCHEIBE AND SCHRÖTER.[91] It was shown to be stable up to temperatures slightly below 150 °C.[91, 97] These findings were also supported by MATHIOT and by LEMKE AND IRMSCHER.[95, 98] The energy level of the CoB pair was tentatively proposed to be at E_C - 0.08 eV.[95]

In addition, Co reacts with hydrogen forming CoH_x complexes. A number of CoH-related complexes were reported by JOST *et al.*[29, 30] and also LEMKE AND IRMSCHER [95]. Six different hydrogen-related complexes (E60, E90, E120, E140, E175, and E200) were observed after wet chemical etching or remote H-plasma treatment in n-type Si whereas only three levels (H50, H90, and H150) were detected in p-type Si.

7.2   Results

DLTS   Figure 7.1 shows DLTS spectra recorded on n- (a) and p-type (b) silicon intentionally doped with cobalt after wet-chemical etching. The dotted spectra are measured close to the surface, while the full-drawn spectra are measured deeper in the bulk. Three prominent peaks $E90_C$, $E140_{Co}$, and $E200_{Co}$ can be observed in the n-type material, whereas one dominant peak H240 and one weak peak H160 are present in p-type material. In order to determine the electrical characteristics of these defects, the Laplace DLTS technique was applied.

After H-plasma treatment, the DLTS spectra differ from those recorded in the etched samples. Figure 7.2 shows the DLTS spectra of n-type silicon. Two different probing regions are shown: close to the surface (dotted line) and deeper in the bulk (full-drawn line). The DLTS spectrum recorded close to the surface shows no signal from $E200_{Co}$, but $E140_{Co}$ and $E90_C$ are still visible. Two new peaks, which were not seen in the etched samples, are also present: a dominant peak at 260K ($E260_C$) and a small peak around 45K ($E45_C$). These two peaks are identical to $E45_C$ and $E260_C$ observed in metal-free silicon presented in Chap. 5. In the bulk, $E45_C$, $E90_C$, and $E260_C$ are no longer visible. Instead, $E200_{Co}$ reappears. $E140_{Co}$ is still observed.

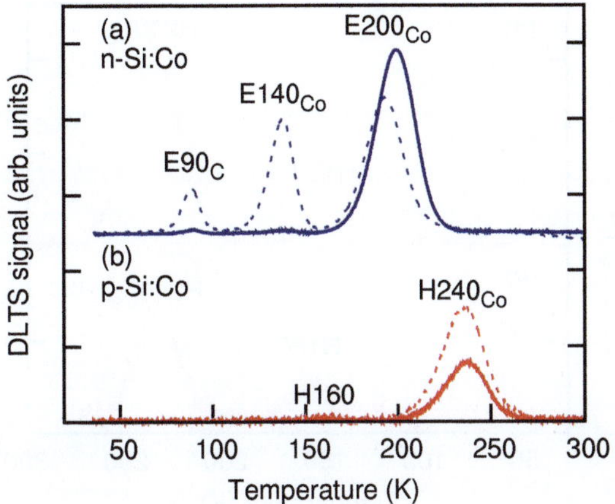

Figure 7.1: DLTS spectra of (a) n-type and (b) p-type Co-doped silicon after wet-chemical etching. The full-drawn lines show spectra measured in the bulk of the sample ($V_R = -8$ V and $V_P = -4$ V), the dotted lines show spectra measured close to the surface ($V_R = -2$ V and $V_P = 0$ V). The spectra were recorded with a filling pulse width of 1 ms and a rate window of 47 s^{-1}.

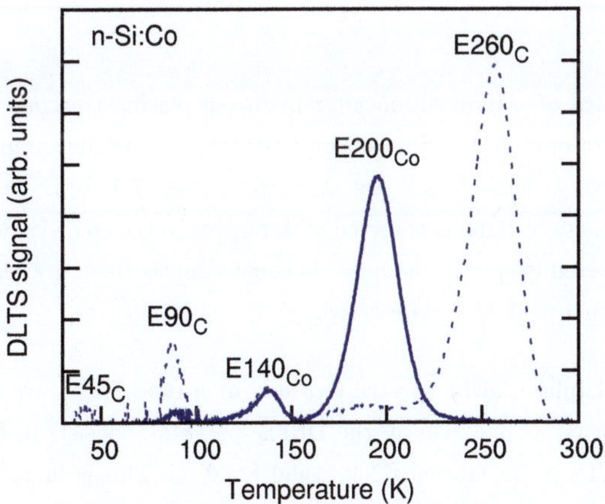

Figure 7.2: DLTS spectra recorded in H-plasma treated n-type silicon doped with cobalt. The dotted lines show the region close to the surface measured with the measurement parameters $V_R = -2V$ and $V_P = 0V$ and the full-drawn lines show the region deeper in the bulk measured with $V_R = -8V$ and $V_P = -4V$. The spectra were taken with a filling pulse width of 1 ms and a rate window of 47 s^{-1}.

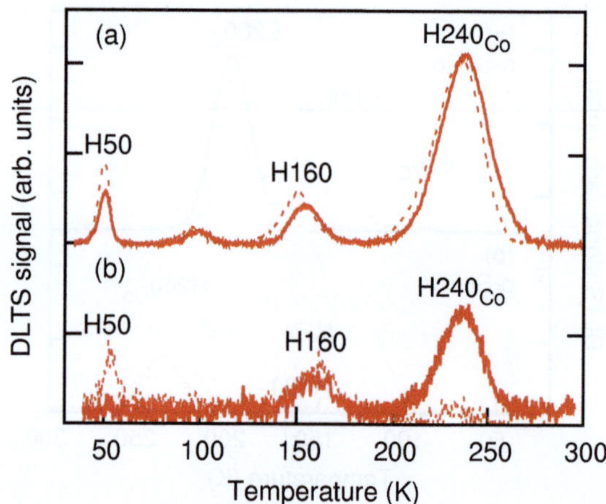

Figure 7.3: DLTS spectra recorded in Co-doped p-type silicon (a) after H-plasma treatment and (b) after reverse bias annealing at 400 K and -5 V (shifted for clarity). The dotted lines show the region close to the surface measured with $V_R = -2V$ and $V_P = 0V$ and the full-drawn lines show the region deeper in the bulk measured with $V_R = -8V$ and $V_P = -4V$. All spectra were taken with a filling pulse width of 1 ms and a rate window of 47 s^{-1}.

The DLTS spectra of p-type silicon after hydrogen plasma treatment are presented in Fig. 7.3 (a). Spectra of etched p-type silicon after reverse bias annealing (RBA) at 400 K with an applied reverse bias of -5 V are also shown (Fig. 7.3 (b)). In both samples, an increase of the intensity of H160 is observed with respect to the etched samples. After RBA, H240$_{Co}$ is only detected deeper in the bulk. In some samples treated with hydrogen plasma or RBA, an additional peak H50 is observed.

Laplace DLTS   Laplace DLTS spectra recorded in n-type silicon are shown in Fig. 7.4. E200$_{Co}$, which appears as one peak in the DLTS spectrum, consists in fact of two closely spaced Laplace DLTS peaks labeled E200$_{Co}$ and E200'$_{Co}$. This is in good agreement with the assumptions made in Ref. [95]. In contrast, only one sharp peak was observed for E90$_C$ and E140$_{Co}$. No additional Laplace DLTS peaks were observed in p-type silicon, where both H240$_{Co}$ and H160 appear as single peaks in the Laplace DLTS spectrum.

Arrhenius plot   The Arrhenius plots for all peaks are shown in Fig. 7.5. The activation energies E_{na} and apparent capture cross sections σ_{na} result from applying Eq. 2.12 and

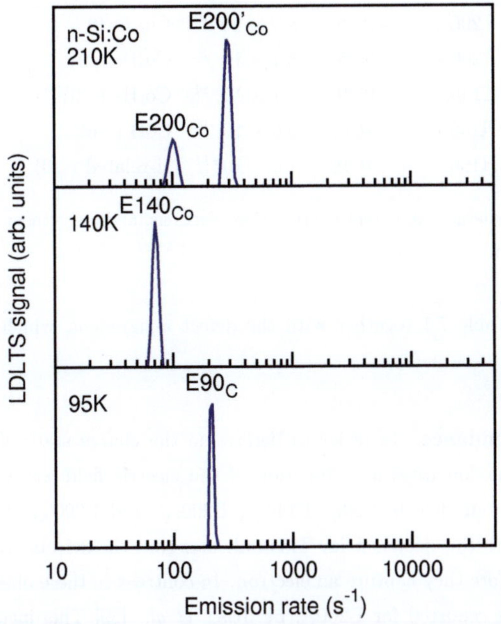

Figure 7.4: Laplace DLTS spectra recorded for E200$_{Co}$, E140$_{Co}$, and E90$_C$.

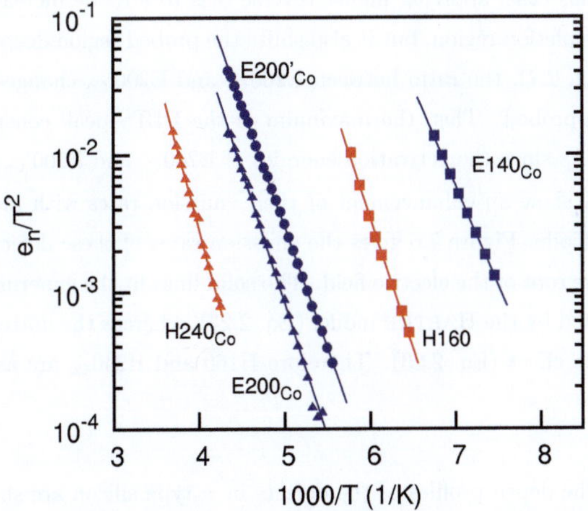

Figure 7.5: Arrhenius plot of all Co-related levels in n- (blue) and p-type (red) silicon.

label	E_{na} (eV)	σ_{na} (cm^2)	assignment
E200$_{Co}$	- 0.39	8.7×10^{-16}	Co$_s$ (-/0)
E200'$_{Co}$	- 0.35	3.7×10^{-16}	Co$_s$H (-/0)
E140$_{Co}$	- 0.29	1.1×10^{-14}	Co$_s$H$_2$ (-/0)
H240$_{Co}$	+ 0.46	9.6×10^{-16}	CoB (-/0)
H160	+ 0.38	1.6×10^{-14}	H-related (-/0)

Table 7.1: Activation energies, apparent capture cross sections, and assignment for all defects seen in n- and p-type silicon.

are summarized in Table 7.1 together with the defect assignment, which is argued in the discussion.

Electric field dependence In order to determine the charge state of the defects, the behavior of their emission rates as a function of the electric field was investigated. The emission rate of all four defects E90$_C$, E140$_{Co}$, E200$_{Co}$, and E200'$_{Co}$ detected in n-type silicon is constant with the electric field. This indicates that the defects are single acceptors which are neutral before they capture an electron. In contrast to these observations, a weak field dependence was reported for E200$_{Co}$ by JOST *et al.* [29] This inconsistence can be explained by the fact that the conventional DLTS technique cannot resolve properly E200$_{Co}$ and E200'$_{Co}$. In Ref. [29] the authors varied the reverse bias while keeping the filling pulse constant. In this case, applying higher reverse bias to a diode increases not only the electric field in the depletion region, but it also shifts the probed region deeper into the bulk. As shown below (Fig. 7.7), the ratio between E200$_{Co}$ and E200'$_{Co}$ changes dramatically if different regions are probed. Then the maximum of the DLTS peak consisting of E200$_{Co}$ and E200'$_{Co}$ also shifts since the activation energies of E200$_{Co}$ and E200'$_{Co}$ are different.

H240$_{Co}$ and H160 show an enhancement of their emission rates with increasing electric field applied to the diode. Figure 7.6 gives the emission rates of these defects recorded as a function of the square root of the electric field. The solid lines fit the experimental data with a fixed slope calculated by the HARTKE model (Eq. 2.24), whereas the dotted lines show the 1D-POOLE-FRENKEL effect (Eq. 2.23). Therefore H160 and H240$_{Co}$ are assigned to single acceptors.

Depth profiles The depth profiles of the defects in n-type silicon are shown in Fig. 7.7. The profile of the hydrogen passivated phosphorous donor [25] is also added to this figure. The local variation of the PH concentration is calculated from the CV-profiles as the differ-

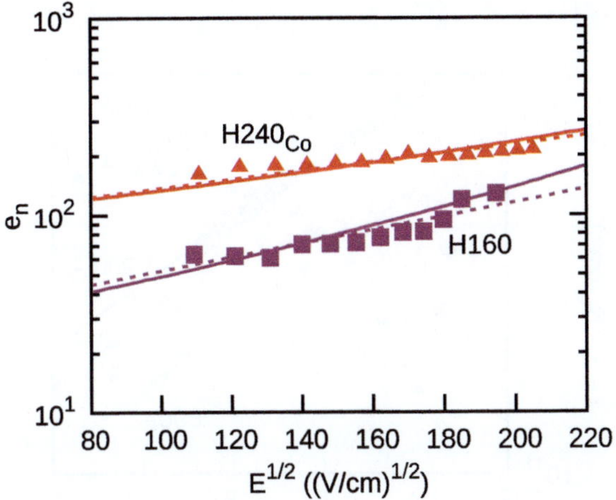

Figure 7.6: Emission rates of H240$_{Co}$ (▲) and H160 (■) versus the square root of the electric field. The solid lines show a fit of the experimental points with the Hartke model (see Eq. 2.24).

ence between the bulk and local free carrier concentration. As shown in Fig. 7.7 three of the four defects, E90$_C$, E140$_{Co}$, and E200'$_{Co}$, can be detected in the region of 2-12 µm and their concentration is significantly reduced when probing deeper in the bulk. In contrast, E200$_{Co}$ shows the opposite behavior. It appears only at a depth of several µm and its concentration increases towards the bulk. E200$_{Co}$ and E200'$_{Co}$ are correlated in the way that the concentration of E200'$_{Co}$ increases in the region where the concentration of E200$_{Co}$ decreases and vice-versa. The concentration of E90$_C$, E140$_{Co}$, and E200'$_{Co}$ increases when more hydrogen is introduced into the sample and can therefore be attributed to hydrogen-related defects.

The depth profiles in Fig. 7.7 were analyzed by comparing the slopes of the concentration tails, as described in Sec. 2.4. This analysis shows that the tails of the hydrogen passivated phosphorous donors PH and of the defects E90$_C$ and E200'$_{Co}$ have identical slopes, whereas the slope of E140$_{Co}$ is twice as steep. E140$_{Co}$ contains therefore twice as much hydrogen as PH, E90$_C$, and E200'$_{Co}$.

The measured depth profiles of the peaks E90$_C$, E140$_{Co}$, E200$_{Co}$, and E200'$_{Co}$ after hydrogen plasma treatment are shown in Fig. 7.8. Only E90$_C$ is observed close to the surface. All other defects appear only deeper in the bulk. This implies a total passivation of these defects by hydrogen close to the surface.

The depth profiles of the hole traps obtained in p-type silicon is shown in Fig. 7.9. In addition, the concentration of the passivated boron acceptors (BH) is plotted as a function

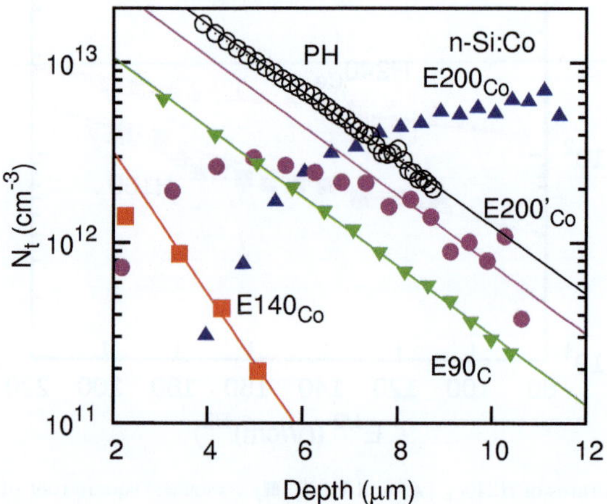

Figure 7.7: Depth profiles of the defects obsesrved in Co-doped n-type silicon. The concentration of the passivated phosphorous donors PH is also presented. The solid lines show the fit of the reduction in concentration of the corresponding defects.

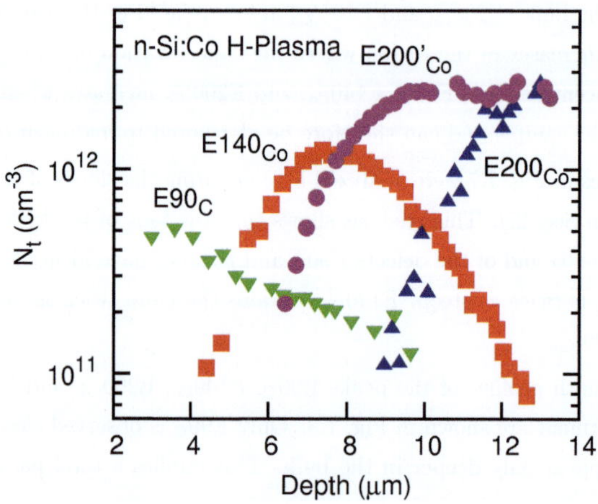

Figure 7.8: Depth profile of the defects seen in Co-doped n-type silicon after plasma hydrogenation.

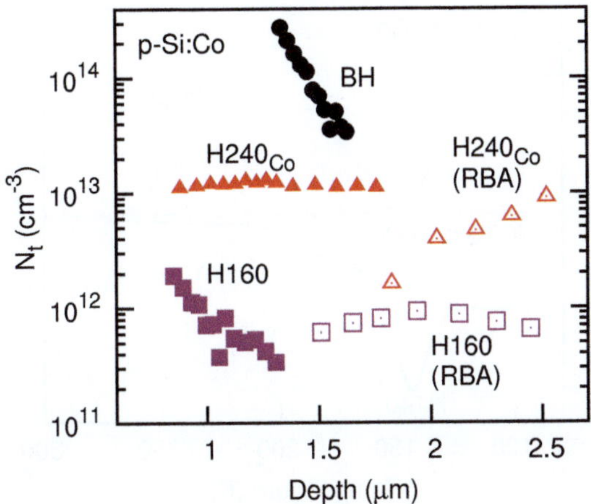

Figure 7.9: Depth profile of the defects observed in Co-doped p-type silicon after etching (closed symbols) and after additional reverse bias annealing (RBA) at 400 K for 20 minutes with -5 V (open symbols). The concentration of the passivated boron acceptors (BH) is also presented.

of depth in the depletion region. The concentration of BH was determined in the same way as that of the passivated shallow donors in Fig. 7.7. Directly after etching, the concentration of H240$_{Co}$ remains constant throughout the depletion region, whereas H160 is present only close to the surface. The concentration of H160 is about one order of magnitude lower than the bulk concentration of H240$_{Co}$.

After an additional RBA of 20 minutes at 400 K with an applied reverse bias of $V_R = -5$ V, H240$_{Co}$ is not observed near the surface. It appears only for depths below 1.8 µm. H160 has a maximum around 2 µm and still has a concentration of about one order of magnitude smaller than H240$_{Co}$.

Hydrogen is known to be positively charged in p-type silicon independently on the position of the Fermi level.[46] During RBA at 400 K, complexes of hydrogen with shallow acceptors are destroyed and free hydrogen located close to the surface drifts in the electric field towards the bulk.[99] There it can bind with shallow acceptors or other crystal imperfections. The depth profile of H160 observed after RBA matches this scenario. During RBA H160 is destroyed near the surface. After the drift of free hydrogen it is formed again deeper in the bulk.

Similarly, after the RBA at 400 K H240 was not observed in the DLTS spectrum at depths below 1.8 µm (see open symbols in Fig. 7.9). This can be interpreted as the drift of positively

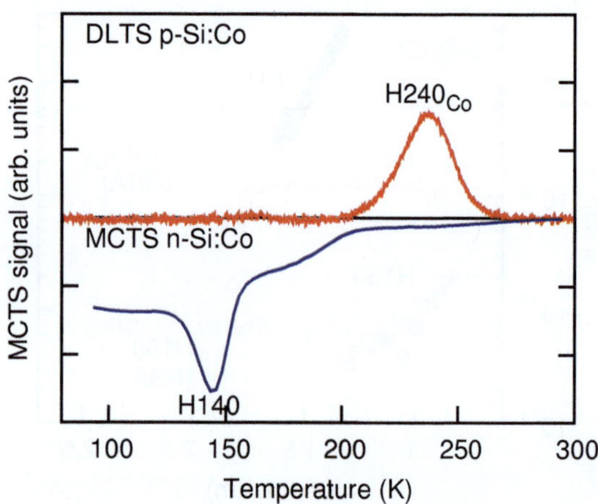

Figure 7.10: MCTS spectrum recorded in Co-doped n-type silicon compared with the DLTS spectrum in Co-doped p-type silicon (same as Fig. 7.1). The black line marks zero. The MCTS spectrum was recorded with the following parameters: $V_R = -2V$, laser pulse width = 1 ms, rate window = 50 s^{-1}.

charged constituents of the H240$_{Co}$ defect towards the bulk of the sample followed by a capture of immobile defects. However, it can be ruled out that H240$_{Co}$ is a hydrogen-related defect since its concentration does not scale with the hydrogen content in the H-plasma treated or etched sample.

MCTS In order to check whether the traps in n-type are also present in the p-type material and vice versa, MCTS measurements were performed in both n- and p-type silicon. The results of these measurements are presented in Figs. 7.10 and 7.11. In n-type silicon we have not found any traces of H240$_{Co}$ in the MCTS spectra. This observation supports that this peak is not correlated with Co$_s$. In the MCTS spectrum of p-type silicon, E200$_{Co}$ is dominant. The presence of this peak in the p-type material indicates that this defect is not correlated with shallow donors and is present in all samples doped with Co. Due to technical limitations, E200$_{Co}$ and E200'$_{Co}$ could not be resolved in the MCTS spectra.

Annealing The thermal stability of the defects was investigated by isochronal annealing steps of 20 min duration. Figure 7.12 shows the normalized concentrations of all observed defects versus the annealing temperature. The hole traps disappear after annealing at 150 °C

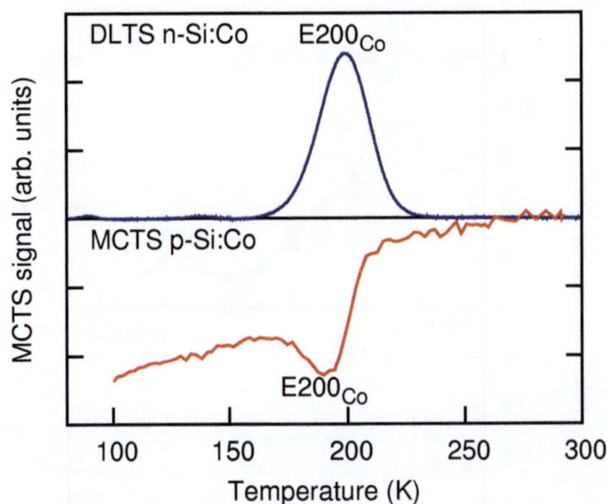

Figure 7.11: MCTS spectrum recorded in Co-doped p-type silicon compared with the DLTS spectrum in Co-doped n-type silicon (same as Fig. 7.1). The black line marks zero. The MCTS spectrum was recorded with the following parameters: $V_R = -4V$, laser pulse width = 1 ms, rate window = 50 s^{-1}.

whereas the electron traps are more stable. The concentration of H160 increases by a factor of 2 at about 100 °C. This observation is consistent with that reported for "Co$_i$" in Ref. [95] and therefore these peaks seem to belong to the same defect. The annealing temperature of E90$_C$ was found to be below 100 °C (not shown in Fig. 7.12). E200'$_{Co}$ and E140$_{Co}$ anneal out at temperatures around 300 °C, whereas E200$_{Co}$ remains present in the samples up to 600 °C. E200'$_{Co}$ doubles in concentration right before annealing out. The concentration of E200$_{Co}$ starts to drop for temperatures higher than 150 °C but does not disappear until annealing at 600 °C.

To illustrate the behavior of E200$_{Co}$ and E200'$_{Co}$ in more detail, the depth profiles of E200$_{Co}$ and E200'$_{Co}$ are shown in Fig. 7.13 directly after etching and after subsequent annealing at 250 °C. Initially, the concentration of E200$_{Co}$ is high in the bulk and drops towards the surface. E200'$_{Co}$ is located closer to the surface and its concentration drops towards the bulk. After annealing, E200$_{Co}$ is pushed further into the bulk. E200'$_{Co}$ has increased in concentration and its maximum has shifted towards the bulk.

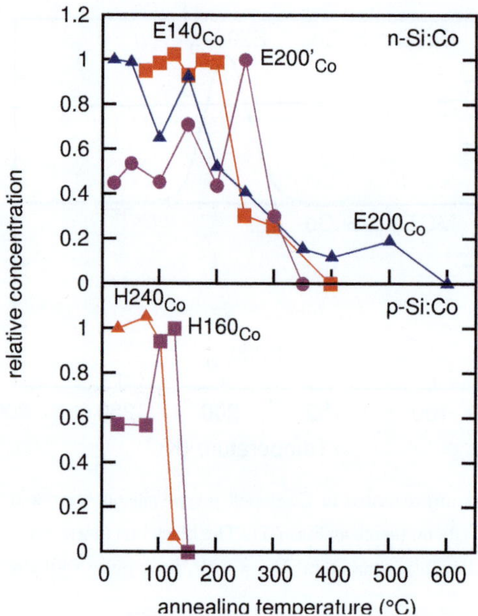

Figure 7.12: The relative concentrations of the levels in (a) n- and (b) p-type silicon versus the annealing temperature. The data point of E200$_{Co}$ at 300 °C is missing because the in-diffusion of hydrogen during the annealing has passivated all E200$_{Co}$ defects in the measurable depth (see text).

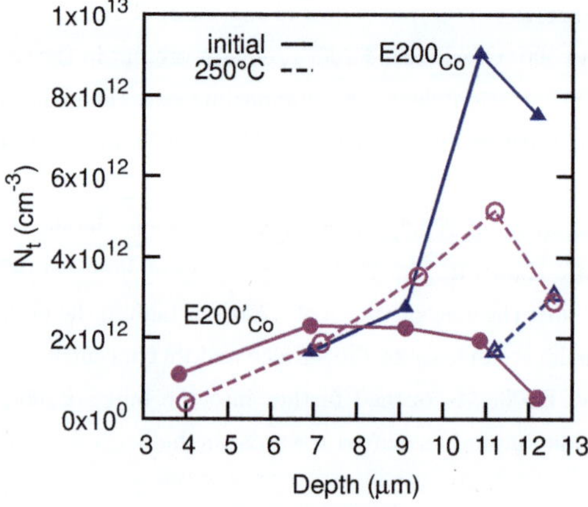

Figure 7.13: Depth profiles of E200$_{Co}$ (▲) and E200'$_{Co}$ (•) in an as-prepared sample (full-drawn lines) and after subsequent annealing at 250 °C (dotted lines).

7.3   Discussion

First, the origin of the peaks previously assigned to Co_s, $E200_{Co}$ and $H240_{Co}$, is discussed. The data presented above rule out that these peaks are two different charge states of the same defect. Both peaks were shown to belong to single acceptors. The different annealing behavior of $H240_{Co}$ and $E200_{Co}$, 150 °C and 700 °C, respectively, further supports the assignment to two different defects. Finally, the absence of $H240_{Co}$ in the MCTS spectrum in n-type silicon is clear evidence against both peaks belonging to the same defect.

In agreement with previous studies, $E200_{Co}$ is assigned to the single acceptor state of Co_s. The annealing temperature of this defect is consistent with that of Co_s as reported in Ref. [12]. In addition, this defect was observed with DLTS and MCTS in n- and p-type silicon, respectively, and it appears only in Co-doped samples.

Since $H240_{Co}$ was shown to belong to a different defect, no other DLTS level which could be correlated with the donor charge state of Co_s was observed. This indicates that Co_s has only an acceptor level in the band gap of silicon or the donor level is too shallow to be observed. The absence of the donor level is in good agreement with the results reported by SCHEIBE AND SCHRÖTER.[91] These authors detected substitutional cobalt by means of Mössbauer spectroscopy but could not observe Co_s-related levels by DLTS in the lower half of the band gap.[91] The observed results also agree well with the data published in Refs. [30, 88].

In contrast to the results of Ref. [95] no deep levels which could be correlated with Co_i in silicon were found. LEMKE AND IRMSCHER suggested that Co_i introduces two levels in the band gap of silicon with electrical properties similar to $E200'_{Co}$ and $H160$. These two defects were shown above to be related to hydrogen. In the depth profiles, $E200'_{Co}$ and $H160$ are located close to the surface, and their intensity correlates with the amount of hydrogen in the sample. In addition, the annealing temperature of these defects was shown to be different. Both $E200'_{Co}$ and $H160$ are acceptor-like defects. These observations exclude that $E200'_{Co}$ and $H160$ belong to different charge states of the same defect. Therefore, Co_i seems to be either not electrically active in silicon or, what is even more likely, Co_i is instable at room temperature due to its high mobility.[92]

If $H240_{Co}$ is not the donor state of Co_s, the question of the origin of this peak needs to be addressed. It seems reasonable to assign $H240_{Co}$ to the single acceptor state of the CoB pair due to the following arguments. $H240_{Co}$ peak cannot be detected by MCTS in n-type silicon since boron is not present there. The annealing temperature of $H240_{Co}$ is similar to that reported by BERGHOLZ[97] and by SCHEIBE AND SCHRÖTER[91] for the CoB pair.

In addition, the RBA experiments show that $H240_{Co}$ consists of positively charged mobile defects which drift in the electric field at about 125 °C and which are not correlated with hydrogen. As mentioned above, Co_i was shown to be highly mobile at 125 °C,[92] which makes Co_i a conceivable constituent of the H240 defect. If this complex is destroyed, Co_i drifts in the electric field towards the bulk where it is again captured by immobile boron.

A number of hydrogen-related defects was also found in both n- and p-type silicon. $E90_C$ can be identified as the CH complex CH_{1BC}. Its properties match those reported in the literature and presented in Chap. 5.

The number of hydrogen atoms involved in the complexes were determined from the slopes of the depth profiles. The slope of $E200'_{Co}$ is identical to that of $E90_C$ and the PH complex. $E200'_{Co}$ is therefore a complex which contains only one hydrogen atom. Since the slope of $E140_{Co}$ is twice as steep as those of $E90_C$, $E200'_{Co}$, and PH, it can be attributed to a complex with two hydrogen atoms.

The concentration of $E140_{Co}$ and $E200'_{Co}$ is inversely proportional to $E200_{Co}$, which is assigned to Co_s. Therefore, $E200'_{Co}$ and $E140_{Co}$ can be identified as substitutional cobalt with one and two hydrogen atoms, respectively.

H160 seems to be not a cobalt related defect and is rather correlated with a complex of hydrogen and an impurity unintentionally introduced during growth. The involvement of hydrogen follows from the increased peak amplitude after hydrogen plasma treatment and after RBA (see Fig. 7.3). Its concentration is about one order of magnitude lower than that of $H240_{Co}$ and no correlation could be found between these defects before and after RBA. Therefore, one can conclude that the intensity of H160 is limited by the concentration of another component involved into this defect.

Similarly, H50 appears in p-samples after plasma hydrogenation or RBA treatment. However, this peak is not always reproducible and therefore most likely belongs to a complex of hydrogen with an unintentionally introduced impurity.

At last, the depth profiles recorded in hydrogenated n-type samples show the reduction of $E140_{Co}$, $E200'_{Co}$, and $E200_{Co}$ towards the surface. Hydrogen plasma treated n-type samples even exhibit a total absence of cobalt-related peaks near the surface (see Fig. 7.8). The passivation of Co-related defects suggests the presence of CoH-related complexes with more than two hydrogen atoms which are electrically inactive. One can speculate that these complexes might consist of three or four hydrogen atoms bound to a Co atom.

Figure 7.14 shows an overview of the levels of cobalt, the cobalt-boron pair, and the cobalt-hydrogen complexes in the band gap of silicon composed from the results of this work. The dashed lines indicate the shift of the levels with hydrogenation.

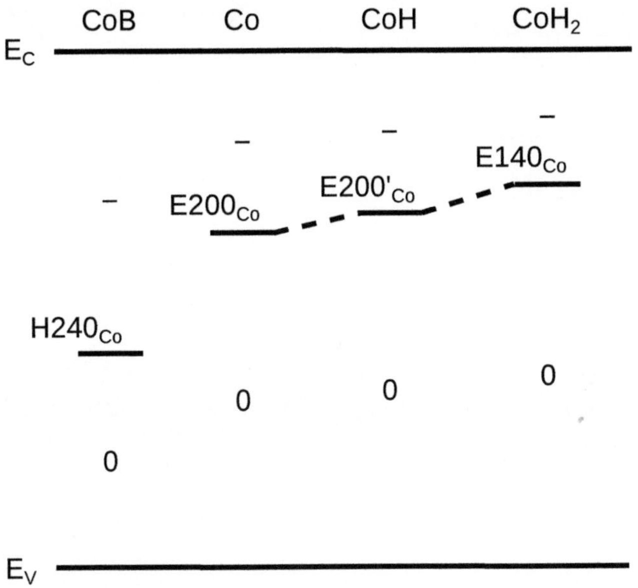

Figure 7.14: Overview of the electric levels of cobalt, the CoB pair, and the CoH complexes in the band gap of silicon with their charge states. The picture is to scale.

7.4 Summary

Substitutional Co_s introduces only a single acceptor level into the upper half of the band gap of silicon. No electrically active levels which can be correlated with interstitial Co_i were observed. The CoB pair can be detected as a deep level with an activation energy of $E_V + 0.4$ eV.

After hydrogenation, Co_s was found to react with hydrogen forming Co_sH-related defects. Using the Laplace DLTS technique, the two deep levels E200'$_{Co}$ and E140$_{Co}$ observed in the upper half of the band gap were concluded to belong to Co_sH and Co_sH$_2$ defects, respectively. Besides the deep levels, the presence of electrically inactive Co_sH complexes was detected.

The level H160 observed in p-type silicon after hydrogenation is assigned to a hydrogen complex with an unintentionally introduced impurity.

Chapter 8

Nickel in silicon

This chapter treats the electrical properties of nickel in silicon and its hydrogen complexes. Different annealing properties in n- and p-type silicon are discussed. From the wealth of hydrogen-related DLTS peaks reported in Ni-doped silicon, four levels belonging to NiH_x complexes can be identified.

8.1   Introduction

Only a small part of the total Ni concentration is electrically active.[12] Two deep levels have been shown by Hall measurements at $E_V + (0.18\text{-}0.25)$ eV and E_C - 0.35 eV.[100] These levels correspond quite well to DLTS results, which show a donor level of nickel at $E_V + (0.15\text{-}0.18)$ eV ($H80_{Ni}$) and an acceptor level at E_C - $(0.38\text{-}0.47)$ eV ($E230_{Ni}$).[101–105] In addition, a third level with an activation energy of E_C - 0.08 eV ($E45_{Ni}$) was reported, which was assigned to the double acceptor of nickel.[32, 104]

The electrically active species is assigned to substitutional nickel.[11, 12, 102, 106] EPR measurements confirm this assignment.[106] This is also in agreement with theoretical calculations, which predict three levels in the band gap for substitutional nickel but no levels for interstitial nickel.[107]

Besides the dominant DLTS peaks assigned to nickel, a number of additional peaks were observed [32, 101–104, 108]. While they are labeled differently in the various publications, they are summarized here with the notation used in this work. In n-type silicon, $E90_{Ni}$, $E150_{Ni}$, and $E270_{Ni}$ are commonly observed. [32, 101–104] In p-type samples, $H280_{Ni}$, $H240_{Ni}$ and $H150_{Ni}$ were often detected.[32, 101–104, 108] Some publications also report on $H50_{Ni}$ and $H190_{Ni}$.[32, 104]

Of these peaks, $E90_{Ni}$, $E270_{Ni}$, $H280_{Ni}$, $H240_{Ni}$, and $H190_{Ni}$ were assigned to NiH complexes.[32, 104] The peaks $H50_{Ni}$, $H150_{Ni}$, and $E150_{Ni}$ were assigned to unknown nickel-hydrogen related complexes. Calculations of the nickel-hydrogen interactions by JONES *et al.* suggest that NiH_2 creates one level around midgap. The hydrogen atoms were proposed to be located in the anti-bonding position.[33] Recent calculations by BACKLUND AND ESTREICHER yield a stable NiH complex with four electrically active levels in the band gap.[27] There is good agreement with JONES *et al.* on the NiH_2 complex, which was also found to be stable in silicon with one level in the band gap at E_C - 0.24 eV.[27] However, the location of the hydrogen atoms at the anti-bonding position was refuted and instead a direct binding to nickel proposed. In addition, calculations show that a reaction of NiH_2 with another hydrogen atom leads to a removal of Ni from the substitutional site and to a partially hydrogenated vacancy, VH_3.[27]

8.2   Results

DLTS The DLTS spectra of wet-chemically etched samples doped with nickel during growth are presented in Fig. 8.1. Figure 8.1 (a) shows the spectra taken in n-type samples, while Fig. 8.1 (b) shows the spectra in p-type samples. In the bulk (full-drawn lines), the peaks $E45_{Ni}$, $E230_{Ni}$, and $H80_{Ni}$ are dominant. Close to the surface (dotted lines), the additional peaks E90, $E270_{Ni}$, $H150_{Ni}$, $H240_{Ni}$, and $H280_{Ni}$ are detected.

The DLTS spectra of nitrogen-doped (see Chap. 4) n-type samples are presented in Fig. 8.2 with (solid lines) and without (dotted lines) in-diffused nickel on both (a) polished and (b) wet-chemically etched samples. $E45_{Ni}$ and $E230_{Ni}$ are dominant in the nickel-containing samples and are not present in the samples without nickel. After etching, the additional peaks E90 and $E270_{Ni}$ appear in the nickel in-diffused samples. In nitrogen doped samples without nickel, a peak $E190_N$ is observed. After etching, the additional peaks $E120_N$ and $E90_C$ appear. The peak position of $E90_C$ is slightly shifted compared to E90 observed in the samples containing nickel.

Figure 8.3 shows the DLTS spectra of nitrogen-doped p-type samples in-diffused with nickel (solid lines) and for comparison spectra of the samples only doped with nitrogen (dotted lines). $H80_{Ni}$ is the dominant peak in the nickel-containing samples. After etching (Fig. 8.3 (b)), the peaks $H150_{Ni}$, $H240_{Ni}$, and $H270_{Ni}$ appear as in samples doped during growth (compare Fig. 8.1). Only one peak $H300_N$ is detected in samples without nickel.

Since $E190_N$, $E120_N$, and $H300_N$ do not appear in nitrogen-doped samples containing nickel, a correlation to nickel can be excluded. DLTS peaks with similar properties as $E190_N$ and $H300_N$ have been reported previously in Refs. [109, 110]. They are attributed to nitrogen-related defects.[71, 109, 110] $E120_N$ appears only after wet-chemical etching, and its intensity correlates with $E190_N$. It is therefore tentatively attributed to a nitrogen-hydrogen complex.

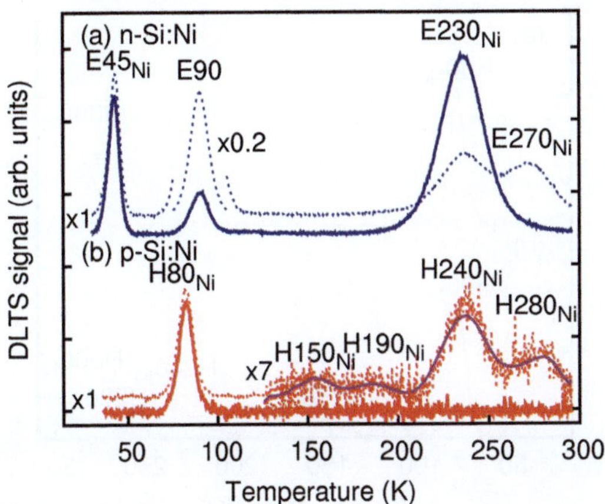

Figure 8.1: DLTS spectra of (a) n-type and (b) p-type silicon doped with nickel during growth after wet chemical etching. The full-drawn lines show the spectra measured in the bulk of the sample ($V_R = -8$ V and $V_P = -4$ V), the dotted lines show the spectra measured close to the surface ($V_R = -2$ V and $V_P = 0$ V). The spectra were measured with a filling pulse width of 3 ms in n-type Si and 1 ms in p-type Si and a rate window of 47 s^{-1}.

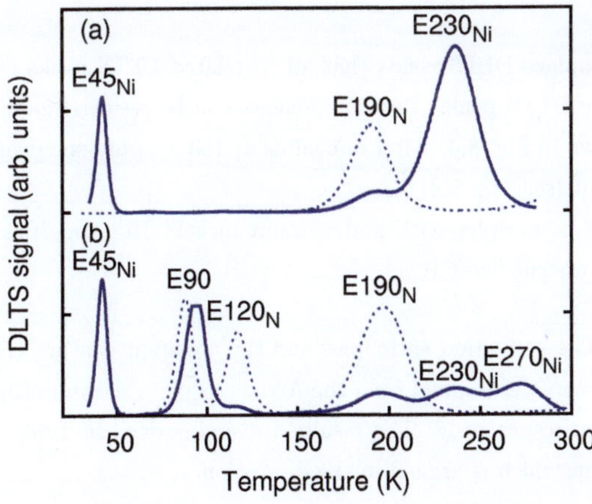

Figure 8.2: DLTS spectra of n-type silicon doped with nitrogen (dotted lines) and in-diffused with nickel (solid lines). Part (a) shows the DLTS spectra on polished samples, part (b) after wet chemical etching. Measurement parameters were $V_R = -2$ V, $V_P = 0$ V, filling pulse width = 3 ms, and rate window = 47 s^{-1} for all spectra.

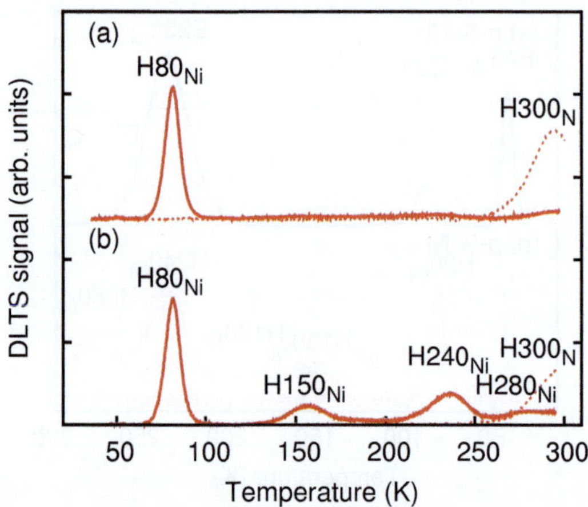

Figure 8.3: DLTS spectra of p-type silicon doped with nitrogen (dotted lines) and in-diffused with nickel (solid lines). Part (a) shows the DLTS spectra on polished samples, part (b) after wet chemical etching. Measurement parameters were $V_R = -2$ V, $V_P = 0$ V, filling pulse width $= 1$ ms, and rate window $= 47$ s^{-1} for all spectra.

Laplace DLTS   Laplace DLTS yields that all Ni-related DLTS peaks except E90 consist of one single Laplace DLTS peak. Two components can be resolved for E90, labeled E90$_C$ and E90$_{Ni}$, as is shown in Fig. 8.4. After annealing at 150 °C, only one peak (E90$_{Ni}$) remains and E90$_C$ anneals out (see Fig. 8.4).

E90$_C$ is the same in samples with and without nickel. Its properties are discussed in Chap. 5 and can be assigned to CH$_{1BC}$.

Arrhenius plot   The activation enthalpies and the apparent capture cross sections of all nickel-related peaks were determined from the Arrhenius plots shown in Figs. 8.5 and 8.6 for n- and p-type silicon, respectively. The results are summarized in Table 8.1 together with the defect assignment, which is argued in the discussion.

Capture cross section   Direct measurements of the capture process of E45$_{Ni}$ at different temperatures are presented in Fig. 8.7, where the normalized difference between measured and saturation peak amplitude is plotted against filling pulse width. From this plot, the sum of the emission and capture rate can be obtained using Eq. 2.19. Since the emission rate

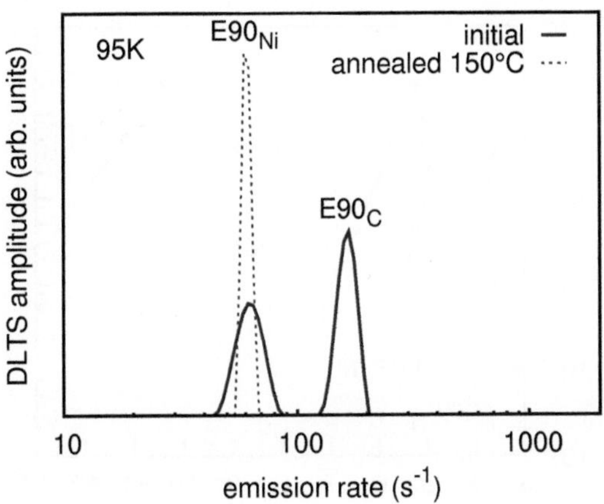

Figure 8.4: Laplace DLTS spectra of n-type silicon doped with nickel after wet chemical etching taken at 95 K. The full-drawn line shows the initial spectrum, the dotted line shows the spectrum measured after annealing the sample at 150 °C for 20 minutes.

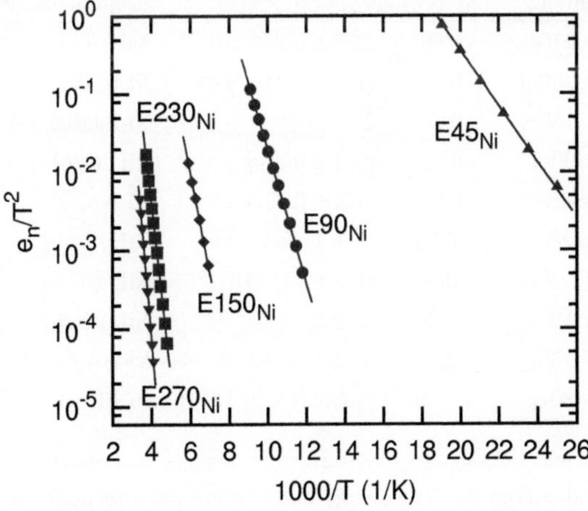

Figure 8.5: Arrhenius plots for all Ni-related levels observed in n-type samples. The corresponding activation enthalpies and capture cross sections are listed in Table 8.1

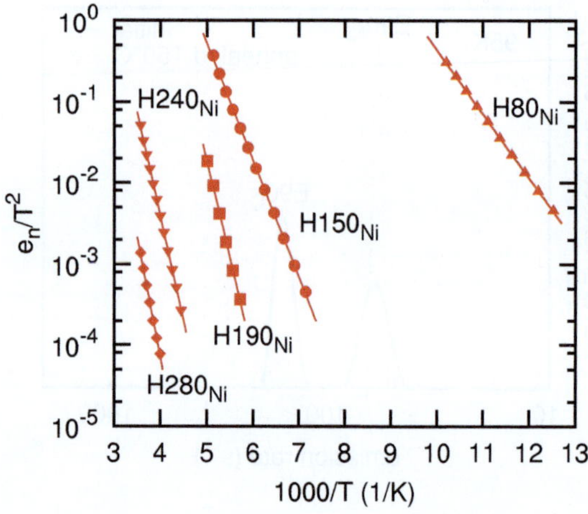

Figure 8.6: Arrhenius plots for all Ni-related levels observed in p-type samples. The corresponding activation enthalpies and capture cross sections are listed in Table 8.1

label	E_{na} (eV)	σ_{na} (cm^2)	assignment
E270$_{Ni}$	-0.60	$(2.7 \pm 0.3) \times 10^{-14}$	NiH$_2$ (-/0)
E230$_{Ni}$	-0.45	$(1.4 \pm 0.1) \times 10^{-15}$	Ni (-/0)
E150$_{Ni}$	-0.26	2×10^{-16}	Ni-related
E90$_{Ni}$	-0.17	$(2.2 \pm 0.8) \times 10^{-15}$	NiH (-/0)
E45$_{Ni}$	-0.07	$(1.2 \pm 0.7) \times 10^{-15}$	Ni (- -/-)
H280$_{Ni}$	$+0.58$	$(2 \pm 1) \times 10^{-14}$	NiH$_2$ (-/0)
H240$_{Ni}$	$+0.49$	$(1.5 \pm 1) \times 10^{-14}$	NiH (0/+)
H190$_{Ni}$	$+0.46$	$(2 \pm 2) \times 10^{-11}$	NiH$_3$ (-/0)
H150$_{Ni}$	$+0.29$	$(4 \pm 1) \times 10^{-15}$	Ni-H-related
H80$_{Ni}$	$+0.16$	$(1.1 \pm 0.3) \times 10^{-14}$	Ni (0/+)

Table 8.1: Activation energies, apparent capture cross sections, and assignment for all Ni-related defects seen in n- and p-type Si. The energy and capture cross sections are mean values over measurements on different samples. The errors indicated for the capture cross section are the statistical standard deviations. The statistical errors for the energies are smaller than the last digit.

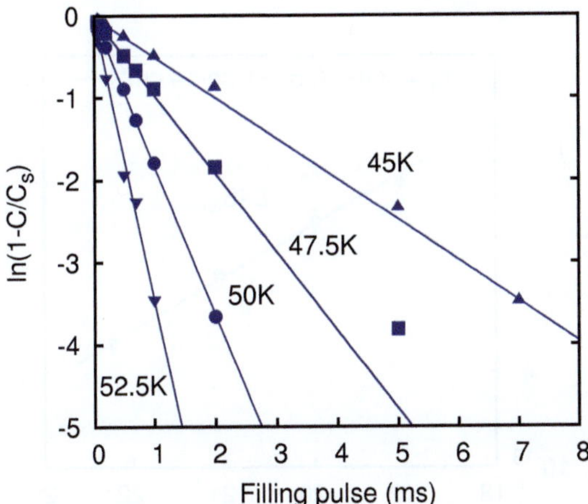

Figure 8.7: Determination of the capture rates of E45$_{Ni}$ at the temperatures indicated. Shown is the logarithm of the difference of the signal amplitude at the corresponding pulse width to the saturation amplitude, normalized to the saturation amplitude. The resulting capture rates are shown in Table 8.2.

is known from the Laplace DLTS measurements, the capture rate can be determined. The capture cross section is then calculated from the capture rate using Eq. 2.2a. The results are given in Table 8.2.

The Arrhenius plot of the measured capture cross sections is shown in Fig. 8.8. From this plot, a barrier height of $E_\sigma = 0.030 \pm 0.005$ eV is determined, which is close to the value of $E_\sigma = (0.04\text{-}0.05)$ eV given in the literature.[32, 104] One should notice, however, that in Refs. [32, 104] the time constant determined from the variation of the filling pulse was directly taken as the capture rate. Since emission and capture rates for this defect are comparable (see Table 8.2), ignoring the emission rate leads to an overestimation of the capture barrier. Therefore, the results presented here can be considered as more precise compared to Refs. [32, 104].

Besides E45$_{Ni}$, also E270$_{Ni}$ exhibits an energy barrier for capture. From the temperature dependence of the measured capture cross sections shown in Fig. 8.9, the barrier is calculated as $E_\sigma = 0.060 \pm 0.005$ eV.

Electric field dependence   Among all peaks observed in Figs. 8.1-8.4 only H190$_{Ni}$ exhibits a dependence of the emission rate on the applied electric field, which is presented in Fig. 8.10.

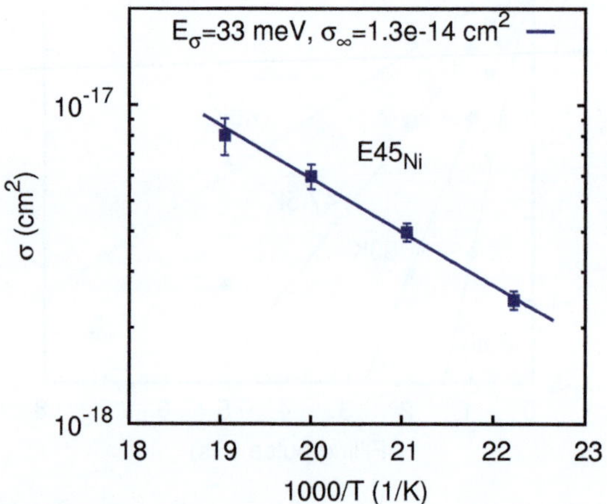

Figure 8.8: Determination of the capture barrier of E45$_{Ni}$. Shown are the capture rates of E45$_{Ni}$ as determined from Fig. 8.7 versus the inverse temperature.

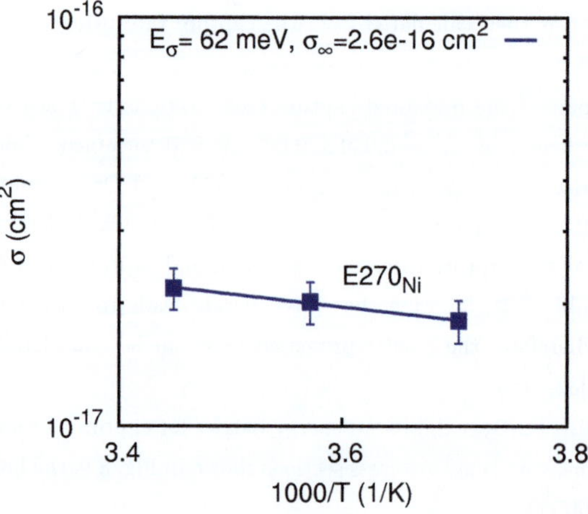

Figure 8.9: Determination of the capture barrier of E270$_{Ni}$ from the directly measured capture rates.

T (K)	$e+c$ (s^{-1})	e (s^{-1})	c (s^{-1})	σ_n (cm^2)
45	490 ± 10	120 ± 5	370 ± 15	$(2.5 \pm 0.2) \times 10^{-18}$
47,5	950 ± 20	330 ± 5	620 ± 25	$(4.0 \pm 0.3) \times 10^{-18}$
50	1810 ± 50	860 ± 10	950 ± 60	$(6.0 \pm 0.5) \times 10^{-18}$
52,5	3450 ± 100	2140 ± 40	1310 ± 140	$(8.0 \pm 1.1) \times 10^{-18}$

Table 8.2: Effective time constant $(e+c)$ of the capture process of E45$_{\mathrm{Ni}}$, its emission and capture rate e and c and the calculated capture cross section σ_n for different temperatures. The time constants were determined from Fig. 8.7.

The theoretical dependencies calculated from the 1D-POOLE-FRENKEL model (dotted line) and from the HARTKE model (solid line) are also plotted. The experimental data is well described by the HARTKE model. H190$_{\mathrm{Ni}}$ is thus assigned to an acceptor level in p-type silicon.

MCTS   In order to determine which of the defects are present in both n- and p-type silicon, MCTS spectra were recorded on n-type silicon samples both with nickel doped during the growth of the crystal and with in-diffused nickel. In Fig. 8.11 (a), the MCTS spectrum of the samples doped during growth is presented. For further analysis, the background,

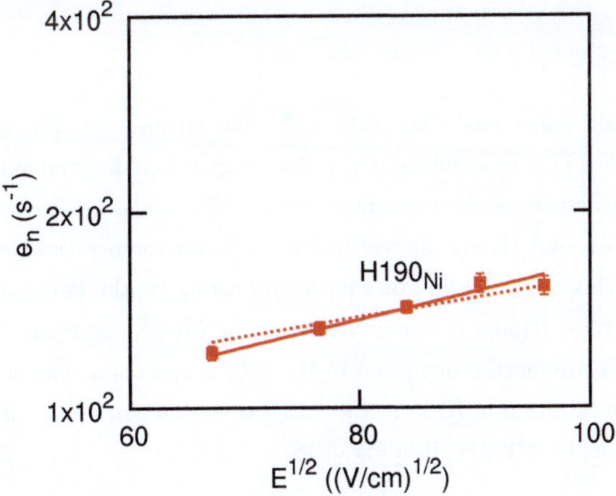

Figure 8.10: Emission rates of H190$_{\mathrm{Ni}}$ versus the square root of the electric field. The solid line shows the dependence calculated by the HARTKE model, the dotted line is calculated from the 1D-POOLE-FRENKEL model (see Sec. 2.3).

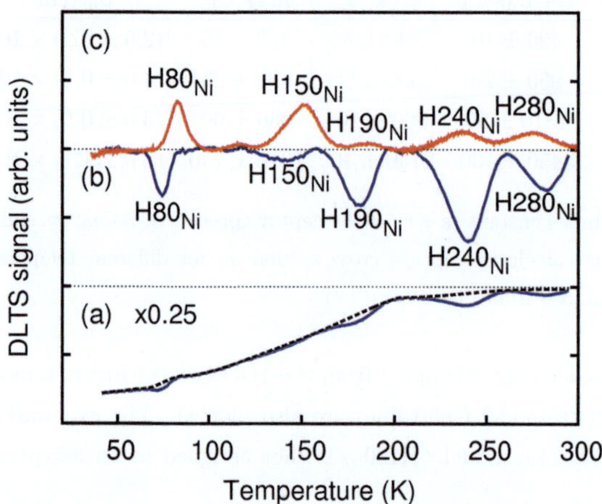

Figure 8.11: (a) MCTS spectrum of n-type silicon doped with nickel during growth after wet chemical etching. The background is marked with the dotted black line. In (b), the background subtracted MCTS spectrum is shown and compared to (c) the DLTS spectrum of the p-type silicon. Measurement conditions were $V_R = -2$ V, $V_P = 0$ V, filling pulse width 1 ms and a rate window of $47\ \mathrm{s}^{-1}$ for the DLTS spectrum. The MCTS spectrum was taken with $V_R = -4$ V, laser pulse width 20 ms, laser power $\sim$90 mW, and a rate window of 80 s^{-1}.

marked in black, was subtracted (Fig. 8.11 (b)). For comparison, Fig. 8.11 (c) shows a DLTS spectrum taken in p-type silicon doped during growth. All dominant peaks observed in DLTS in p-type samples can be reproduced using MCTS on n-type samples. The MCTS peaks H80$_{Ni}$, H150$_{Ni}$, and H190$_{Ni}$ appear shifted to lower temperatures compared to the DLTS spectrum. This might be due to additional heating by the laser pulse. In contrast to the DLTS spectrum, H150$_{Ni}$ is barely visible in the MCTS spectrum. H190$_{Ni}$, H240$_{Ni}$, and H280$_{Ni}$ however are more pronounced in the MCTS spectrum compared to the DLTS spectrum. The concentration of H190 in the DLTS spectrum can be significantly increased by annealing the sample at 120 °C (see Fig. 8.16).

The MCTS spectrum of the in-diffused samples is presented in Fig. 8.12. Here, H150$_{Ni}$ can not be detected in the MCTS spectrum. As in the samples doped during growth, H190$_{Ni}$, H240$_{Ni}$, and H280$_{Ni}$ are more pronounced in the MCTS spectrum than in the DLTS spectrum. In addition, an unidentified feature appears at $\sim$ 100 K in the MCTS spectrum.

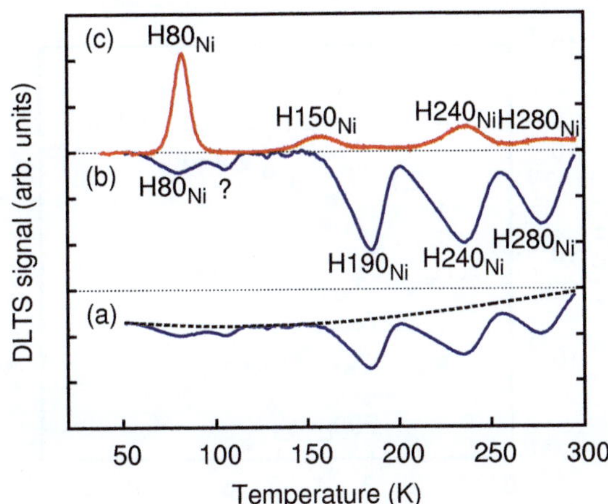

Figure 8.12: (a) MCTS spectrum of n-type silicon in-diffused with nickel after wet-chemical etching. The background is marked with the dotted black line. In (b), the background subtracted MCTS spectrum is shown and compared to (c) the DLTS spectrum of the p-type silicon. Measurement conditions were $V_R = -2$ V, $V_P = 0$ V, filling pulse width 1 ms and a rate window of 47 s^{-1} for the DLTS spectrum. The MCTS spectrum was taken with $V_R = -4$ V, laser pulse width 20 ms, laser power $\sim$90 mW and a rate window of 80 s^{-1}.

Depth profiles   Figure 8.13 shows the concentration depth profiles of the defects observed in n-type silicon. The concentration of E45 was corrected by the factor $(c + e)/c$ (see Eq. 2.18). The depth profiles of E45$_{Ni}$ and E230$_{Ni}$ are almost identical and drop towards the surface. This indicates a passivation of the defects by hydrogen.

E90$_C$, E90$_{Ni}$, and E280$_{Ni}$ are located close to the surface. Their concentration increases when a higher amount of hydrogen is introduced into the samples. Therefore, they can be attributed to hydrogen-related defects. Using the method described in Sec. 2.4, one obtains a ratio of 1:1:2 for the slopes of the reduction of concentration of E90$_C$, E90$_{Ni}$, and E280$_{Ni}$.

The concentration depth profiles of the defects observed in p-type silicon after wet-chemical etching are shown in Fig. 8.14. Besides the concentration of the DLTS peaks H80$_{Ni}$, H280$_{Ni}$, H240$_{Ni}$, and H190$_{Ni}$, the concentration of the passivated boron acceptors (BH) is also shown. It was calculated from the CV profile as the difference between the measured carrier concentration and the saturation concentration deep in the bulk. No reliable depth profile of H150$_{Ni}$ could be obtained, as its concentration varied between different samples.

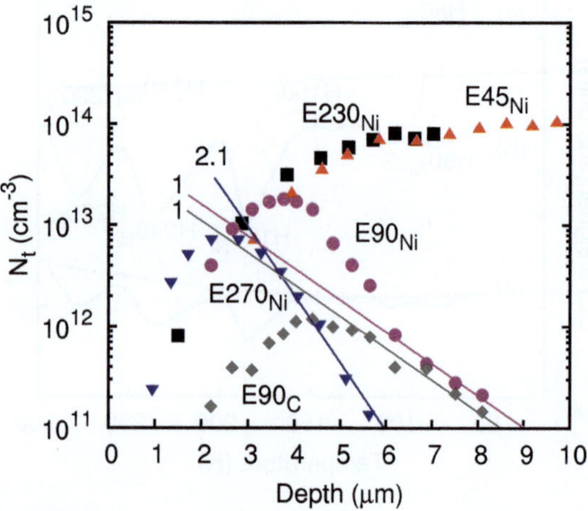

Figure 8.13: Depth profile of the defects observed in n-type Si after etching. The fits to the concentration tails and their slopes normalized to the slope of E90$_C$ are shown as solid lines.

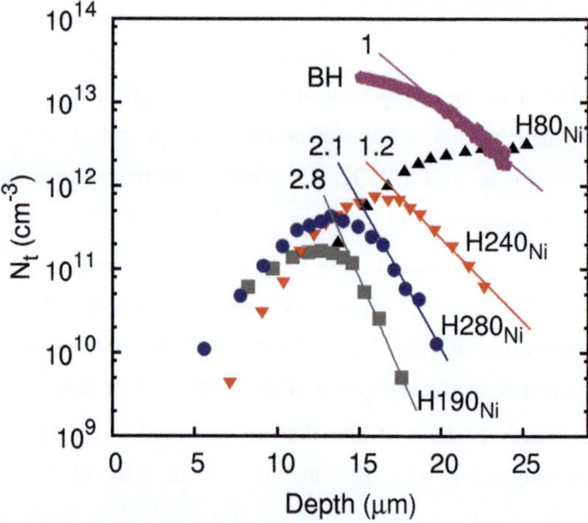

Figure 8.14: Depth profile of the defects observed in p-type Si after etching and of the concentration of the passivated boron acceptors (BH). The fits to the concentration tails and their slopes normalized to the slope of BH are shown as solid lines.

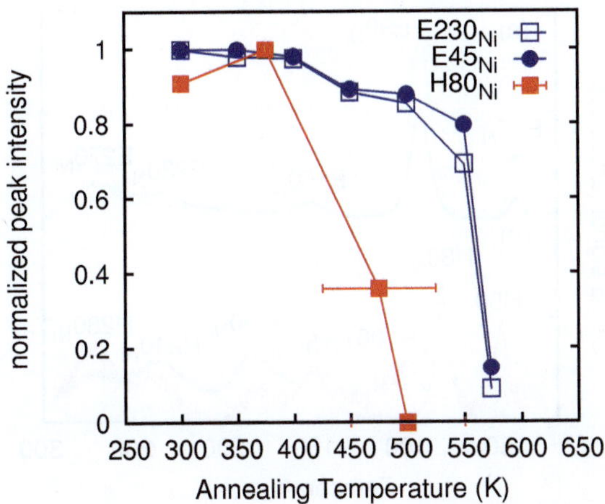

Figure 8.15: The relative concentrations of the levels attributed to nickel versus the annealing temperature.

Since the BH-complex contains one hydrogen atom, its slope is used as a reference when comparing the reduction of the concentration. One then obtains a ratio of $\sim$1:2:3 for H240$_{Ni}$:H280$_{Ni}$:H190$_{Ni}$.

Annealing   Figure 8.15 shows the annealing behavior of the peaks E45$_{Ni}$, E230$_{Ni}$, and H80$_{Ni}$. E45$_{Ni}$ and E230$_{Ni}$ follow each other closely and disappear after annealing at 300 °C. H80$_{Ni}$ anneals at temperatures around 200 - 250 °C.

DLTS spectra of annealed samples are presented in Fig. 8.16. In p-type samples, annealing at 120 °C increases the intensities of H190$_{Ni}$, H240$_{Ni}$, and H280$_{Ni}$. In addition, the DLTS peaks H50 and H100 appear in samples doped during growth. Annealing at about 200 °C leads to the appearance of a broad peak around 220 K in some p-type samples. In n-type samples, E150$_{Ni}$ is observed after annealing at 150 °C.

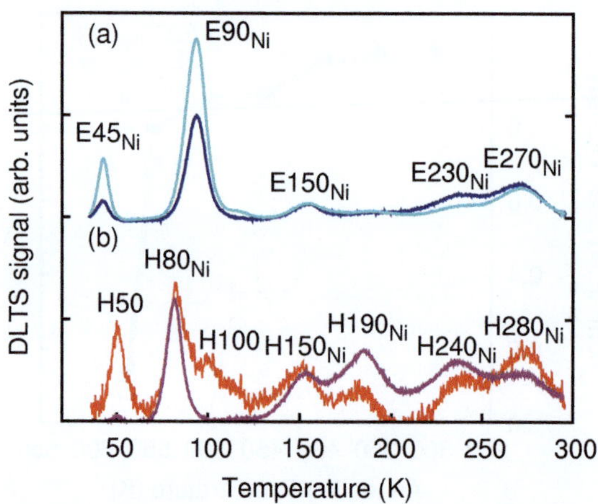

Figure 8.16: DLTS spectra of nickel-doped samples after wet-chemical etching and additional annealing. Part (a) shows n-type samples with nickel added during growth (blue line) and with in-diffused nickel (light blue line) after 150 °C annealing. In (b), p-type samples are shown after annealing at 120 °C, with the samples doped during growth in red and the in-diffused samples in magenta. The spectra were measured with $V_R = -2$ V, $V_P = 0$ V, a filling pulse width of 1 ms, and a rate window of 47 s^{-1}.

8.3   Discussion

Three levels in the band gap of silicon are attributed to substitutional nickel: the double acceptor level at E_C - 0.07 eV (E45$_{\text{Ni}}$), the acceptor level at E_C - 0.45 eV (E230$_{\text{Ni}}$), and the donor level at $E_V + 0.15$ eV (H80$_{\text{Ni}}$) [100–105].

The concentration profiles of E45$_{\text{Ni}}$ and E230$_{\text{Ni}}$ are identical in all samples, and their annealing behavior is very similar. This is in good agreement to the literature assignment of these peaks to two different charge states of substitutional nickel.[100–105] The small capture cross section of E45$_{\text{Ni}}$ ($< 10^{-17}$ cm^2) is in agreement with the assignment to the double acceptor. No field effect is visible for either E45$_{\text{Ni}}$ or E230$_{\text{Ni}}$, but the capture barrier of E45$_{\text{Ni}}$ can be responsible for suppressing the POOLE-FRENKEL effect.[111] Therefore, E45$_{\text{Ni}}$ and E230$_{\text{Ni}}$ can be attributed to the double and single acceptor of substitutional nickel, respectively.

H80$_{\text{Ni}}$ anneals at lower temperatures than the nickel peaks in n-type silicon. However, H80$_{\text{Ni}}$ was always observed in MCTS together with E45$_{\text{Ni}}$ and E230$_{\text{Ni}}$ in both in-diffused samples and in samples doped during growth. In addition, H80$_{\text{Ni}}$ appears in the MCTS spectra of n-type samples even after annealing at 225 °C, a temperature which removes H80$_{Ni}$ from p-type samples. Therefore, different annealing pathways in n- and p-type silicon seem more reasonable than an assignment of H80$_{Ni}$ to a different defect than substitutional nickel.

At an annealing temperature of 200 °C, the Fermi level lies close to midgap, so that the charge state of nickel should be the same in both n- and p-type silicon. This makes a different annealing behavior of H80$_{Ni}$ and E45$_{Ni}$/E230$_{Ni}$ induced by different charge states of substitutional nickel very unlikely. The influence of hydrogen on the annealing behavior of these defects can also be ruled out, since no difference in the annealing behavior of H80$_{Ni}$ was observed between cleaved and etched samples. Also, the hydrogen related peaks in p-type silicon disappear at the same temperature as H80$_{Ni}$ and no recovery of H80$_{Ni}$ can be observed.

The lower annealing temperature of H80$_{Ni}$ can only be explained by different concentrations of extended defects or nickel precipitates. These could modify the crystal lattice in the neighborhood of nickel in n- and p-type silicon. Evidence for the presence of precipitates is the appearance of a broad peak around 220K after annealing at 200 °C in some p-type samples. This peak is similar to those reported for NiSi$_2$ in p-type silicon.[41] The concentration and size of NiSi$_2$ was shown to depend on the dislocation density [112] and is influenced by annealing at temperatures as low as 100 °C [41, 112, 113]. This supports the idea that nickel precipitation affects the annealing temperature of substitutional nickel differently in n- and p-type silicon.

After etching, a number of additional peaks appear in the DLTS spectra (see Figs. 8.1, 8.2, 8.3). Some of these peaks have been reported previously and were assigned to NiH-complexes.[32, 104] However, the results presented here warrant a reassessment of their origins.

The DLTS peak E90 consists of two components, as is shown in Fig. 8.4. The presence of two components of E90 has already been suggested by SACHSE.[32] However, in this work a direct observation of the two components using Laplace DLTS was possible. The electrical and annealing properties of E90$_C$ match well those presented in Chap. 5. It can therefore be attributed to CH$_{1BC}$. The second component E90$_{Ni}$ is more stable compared to E90$_C$ and appears only in nickel-doped samples. Its concentration depends on the amount of hydrogen inside the sample.

The slope of the concentration depth profile of E90$_{Ni}$ is identical to that of E90$_C$. Since it is known that CH contains only one hydrogen atom, E90$_{Ni}$ can be assigned to the single acceptor state of NiH. This is consistent with the absence of the POOLE-FRENKEL effect. The assignment to NiH is in disagreement with the results reported in Refs. [32, 104] where

E90$_{Ni}$ was tentatively attributed to NiH$_2$. However, in the previous works the concentration depth profiles of E90$_C$ and E90$_{Ni}$ could not be well resolved since the conventional DLTS technique was used. Together with relatively low nickel concentrations, this could lead to a misinterpretation of the origin of E90$_{Ni}$.

The slope of the second NiH-complex, E270$_{Ni}$, shows twice the steepness of E90$_C$, and is therefore attributed to the single acceptor state of NiH$_2$. This is again inconsistent with the assignment in Refs. [32, 104]. However, as mentioned above, this can also be a result of the comparison of depth profiles of E270$_{Ni}$ and E90$_{Ni}$.

In the p-type samples, the three peaks H150$_{Ni}$, H240$_{Ni}$, and H280$_{Ni}$ appear directly after etching in both types of samples, doped during growth and in-diffused (see Figs. 8.1 and 8.3). Additionally, H190$_{Ni}$ was observed in samples doped with nickel during growth. Among these four peaks, H150$_{Ni}$ does not appear in the MCTS spectra of in-diffused n-type samples (Fig. 8.12) and only in very small concentrations in the MCTS spectra of samples doped during growth (Fig. 8.11). Furthermore, DLTS measurements on different samples show varying concentrations of this defect. Therefore, H150$_{Ni}$ is attributed to a more complex Ni-related defect compared to the NiH$_x$-complexes.

H190$_{Ni}$ was observed prominently in the MCTS spectra in all samples directly after etching. It is dominant close to the surface and diminishes quickly when measuring deeper in the bulk. The concentration of this defect increases in p-type silicon after annealing at 120 °C regardless of the method of nickel introduction. The strong increase in the DLTS spectra after annealing can be explained by a drift of hydrogen, which is released from BH or other H-related defects, in the intrinsic field of the Schottky diode from the surface towards the bulk.[20]

As mentioned above, an analysis of the depth profile of H190$_{Ni}$ shows that this defect contains 3 times more hydrogen atoms than the BH complex. Therefore, H190$_{Ni}$ is assigned to the single acceptor state of NiH$_3$.

The analysis of the concentration depth profile of H240$_{Ni}$ allows to assign this defect to NiH. The absence of the POOLE-FRENKEL effect suggests that H240$_{Ni}$ belongs to a single donor.

By comparing H280$_{Ni}$ and E270$_{Ni}$, some similarities are apparent. E270$_{Ni}$ introduces a level at about 0.54 eV below the conduction band whereas the energy position of H280$_{Ni}$ was determined to about 0.58 eV above the valence band. Since the band gap of silicon should be around 1.114 eV at room temperature, these traps have, within an experimental error, the same position in the band gap. Both peaks are attributed to complexes with two hydrogen atoms. It is then very likely that these peaks belong to the same defect level which can be observed both in n- and p-type silicon. This defect is attributed to the single acceptor state

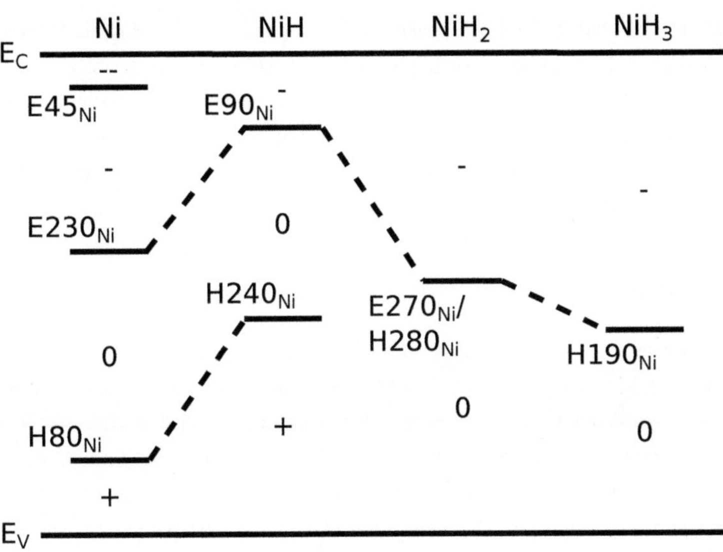

Figure 8.17: Overview of the electric levels of nickel and the NiH complexes in the band gap of silicon with their charge states. The picture is to scale.

of NiH$_2$, even though no electric field dependence of the emission rate of H280$_{Ni}$ has been observed. The low amplitude of H280$_{Ni}$ and a weak field dependence as obtained from the HARTKE model at these temperatures (see Sec. 2.3) might explain why no POOLE-FRENKEL effect could be detected.

The total nickel concentration in the bulk is higher than sum of the nickel and nickel-hydrogen concentrations close to the surface. This suggests the existence of an electrically inactive NiH-complex, probably NiH$_4$, besides the electrically active levels of NiH, NiH$_2$, and NiH$_3$. This is consistent with the results of SACHSE.[32] Alternatively, as was predicted by theory, the reaction of hydrogen with nickel might result in a removal of nickel from the substitutional site and create a partially hydrogenated vacancy.[27] DLTS peaks were reported for several vacancy-hydrogen complexes [53, 114, 115], whose presence might be expected after passivation. However, none of these peaks could be observed in any of the samples.

Figure 8.17 shows an overview of the levels of nickel and the nickel-hydrogen complexes in the band gap of silicon composed from the results of this work. The dashed lines show the shift of the level positions upon hydrogenation.

Besides the Ni- and NiH-related peaks the minor peaks $E150_{Ni}$, H50, and H100 appear after annealing of the samples above 100 °C. H50 and H100 have significantly smaller intensities compared to the Ni-related peaks and therefore they can be attributed to some other unknown impurities which are present in the material. In contrast, $E150_{Ni}$ appears in all samples investigated and doped with nickel. Therefore, it can be linked with a Ni-related defect, but its structure remains unclear.

8.4   Summary

The existence of three levels of substitutional nickel, $E45_{Ni}$, $E230_{Ni}$, and $H80_{Ni}$, was confirmed. The different annealing behavior of the peaks in n- and p-type silicon was suggested to be caused by different concentrations of dislocations and/or nickel precipitates. The single acceptor and single donor states of NiH were found to introduce levels at about E_C - 0.17 eV and E_V + 0.49 eV, respectively. The presence of the single acceptor states of NiH_2 and NiH_3 was observed in p-type silicon. These levels were found to be located close to the midgap at about 0.58 eV and 0.46 eV above the valence band.

A total passivation of nickel by hydrogen was observed. Two possible models were presented: (i) the formation of an electrically inactive NiH_4 complex or (ii) the removal of nickel from the substitutional site and the creation of a partially hydrogenated vacancy [27]. The absence of DLTS peaks related to vacancy-hydrogen complexes favors the first model.

Chapter 9

Conclusion

In this thesis, a comprehensive study on the electrical properties of the three transition metals cobalt, titanium, and nickel in silicon, and of their hydrogen complexes is presented. Here, these metals are compared with each other and with neighboring elements from the periodic table.

Cobalt and nickel are expected to behave similarly in silicon. They are both electrically active in the substitutional species, and are neighbors in the periodic table (^{27}Co and ^{28}Ni). Theory proposes that for a substitutional $3d$-TM in a covalent semiconductor, the $3d$ orbital reacts with the states of the vacancy created by the dangling bonds.[116] In this model, increasing the amount of electrons in the d-orbital (i.e. moving along a row of the periodic table) leads to a continuous shift of the deep levels in the band gap.

Following this idea, Fig. 9.1 shows a comparison of the levels of the substitutional species of iron, cobalt, nickel, and copper. These four elements are neighbors in the periodic table. The data for iron and copper is taken from Ref. [117] and Refs. [13, 118], respectively. It should be noted that the acceptor level of substitutional iron is still under debate. The existence of substitutional iron is known from Mössbauer spectroscopy and from channeling data.[119, 120] Theory calculates that substitutional iron should either introduce no level in the band gap [94, 121] or one acceptor level [107]. Experimentally, a level at E_C - 0.38 eV was tentatively attributed to substitutional iron.[117] The dominant electrical activity of iron, however, results from interstitial iron.[9]

Iron and cobalt only exhibit one acceptor level in the band gap, whereas nickel and copper both have a double acceptor, a single acceptor, and a donor level. The acceptor level shifts upwards from copper towards lower d-shell occupation. Similarly, the double acceptor shifts upwards from copper to nickel and the donor level downwards. Following this trend, one can explain the absence of the double acceptor and single donor levels in cobalt and iron as a crossing of the level position into the conduction and valence band, respectively.

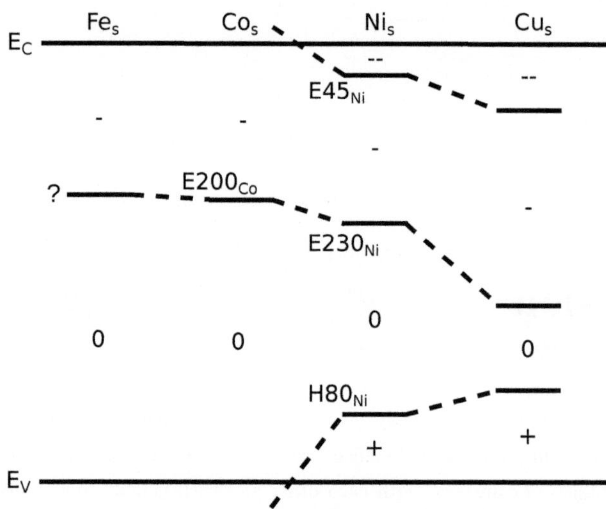

Figure 9.1: An overview of the substitutional levels of cobalt and nickel and their neighbors in the periodic table, iron and copper. The data for iron was taken from Ref. [117], the data for copper from Refs. [13, 118].

Looking at the formation of metal-acceptor pairs for these neighboring elements, the NiB pair is missing. Iron forms the well-known FeB pair with a level at $E_V + 0.10$ eV.[9] Studies of the CoB and other cobalt-acceptor pairs were published previously [91, 95, 97, 98] and the CoB pair was confirmed in this work. Copper is known to create electrically inactive CuB pairs.[122] Since the level of the TMB pair shifts upwards in the band gap of silicon from FeB ($E_V + 0.10$ eV [9]) to CoB ($E_V + 0.46$ eV), the NiB pair might create a level in the upper half of the band gap if this shift continues. This would also be consistent with the electrically inactive CuB pair, which might have a level in the conduction band. However, attempted MCTS measurements of p-type silicon containing nickel were unsuccessful. Future studies might resolve this puzzle of the missing NiB pair.

Upon hydrogenation, the TM levels in the band gap are shifted, creating new electrically active levels in the band gap. It was previously observed that the pattern of this hydrogenation shift is remarkably similar within one group of the periodic table.[13, 128] Therefore, Fig. 9.2 compares the hydrogenation patterns of cobalt and nickel within their respective group. The data for the $4d$ and $5d$ elements is taken from the literature, namely from Ref. [123] (Rh), Ref. [124] (Ir), Refs. [32, 125] (Pd), and Refs. [32, 126, 127] (Pt). It should

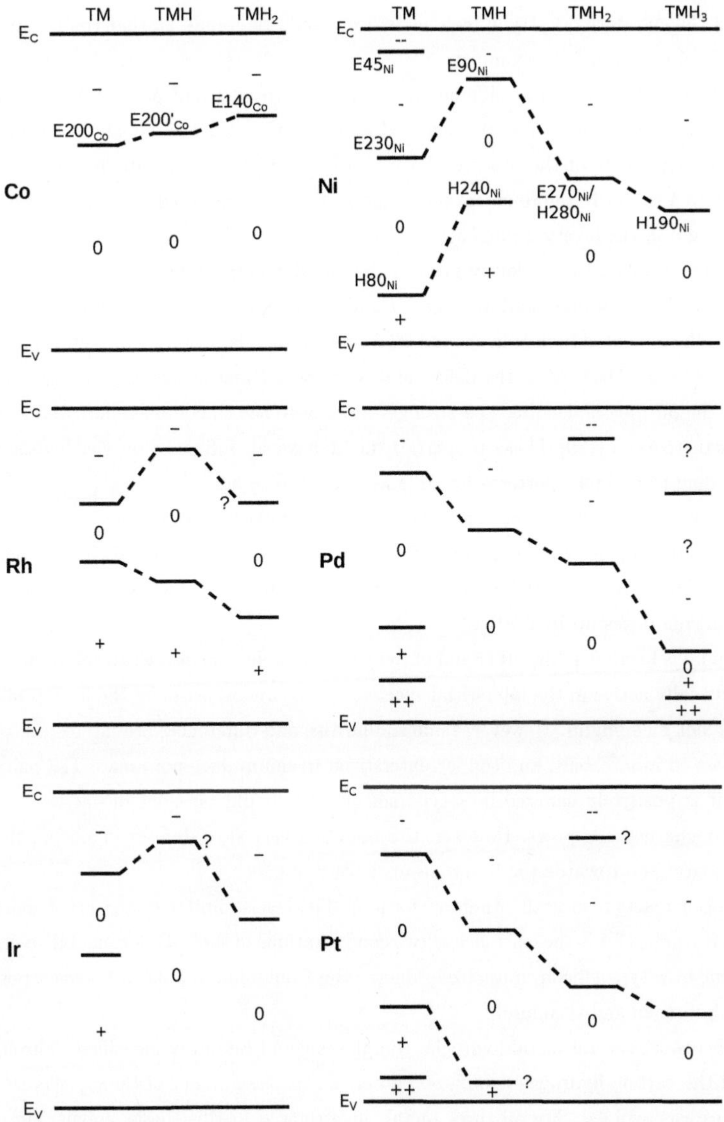

Figure 9.2: A comparison of the levels of Co and Ni and the CoH (NiH) complexes with the elements of the same group, rhodium and iridium (below cobalt) and palladium and platinum (below nickel). The data was taken from Ref. [123] (Rh), Ref. [124] (Ir), Refs. [32, 125] (Pd), and Refs. [32, 126, 127] (Pt).

be noted that no data on p-type Ir-doped silicon could be found, so that the picture of the iridium level in Fig. 9.2 is incomplete.

The $3d$ metals show quite different patterns than the $4d$ and $5d$ metals. In part this might be due to the fact that the renewed analysis of nickel and cobalt in silicon led to several reassignments of the observed levels. Subjecting rhodium, iridium, palladium, and platinum to a similarly rigorous reexamination as the one conducted here might also lead to some changes in the level assignment.

However, it was also previously noticed that fundamental differences exist between the $3d$ TMs in silicon on one hand and the $4d$ and $5d$ TMs on the other hand. BEELER AND SCHEFFLER calcualted high-spin ground states for $3d$ metals, whereas $4d$ metals have a low-spin ground state.[129] Also, the defect states of the $4d$ metals are more delocalized than those of the $3d$ metals, leading to a stronger binding to the silicon neighbors and a reduced local magnetization.[129] These properties might have an influence on the hydrogenation shift, leading to different patterns for $3d$ and $4d$ metals in silicon.

Regardless of the different patterns observed, one similarity within a group of the periodic table is striking: the maximum number of hydrogen atoms in an electrically active complex is the same. Group 10 elements form active complexes up to TMH_3, whereas group 9 elements only form complexes up to TMH_2.

Titanium is located at the other end of the periodic table. In contrast to cobalt and nickel, it is electrically active in the interstitial species. A direct comparison of the level positions is therefore not meaningful. However, some similarities and differences are interesting to note. In contrast to iron, cobalt, and copper, interstitial titanium does not create TiB pairs, even though it is positively charged in p-type silicon and should be Coulomb attracted to the negatively charged acceptors. However, titanium is a very slow diffuser in silicon, therefore a drift of titanium towards the boron atoms is very unlikely.

Hydrogen reacts also with titanium, forming TiH-levels similar to the other metals. A singular feature so far is the existence of two configurations of the TiH-defect. TiH-complexes are absent in p-type silicon, most likely due to the Coulombic repulsion between positively charged hydrogen and titanium.

In conclusion, several metal-hydrogen complexes could be newly identified. Through the study of the carbon-hydrogen complexes, an erroneous assignment of their peaks to TM-H complexes was avoided. For all three metals investigated in this thesis, cobalt, nickel, and titanium, a total passivation by hydrogen can be achieved if the hydrogen concentration is sufficiently high. A comparison of the observed levels with the neighbors in the periodic table shows a trend of the level positions for elements within a row, but differences between $3d$ TMs on one hand and the $4d$ and $5d$ TMs on the other hand.

Bibliography

[1] G. Coletti, P. C. P. Bronsveld, G. Hahn, W. Warta, D. Macdonald, B. Ceccaroli, K. Wambach, N. Le Quang, and J. M. Fernandez, *Impact of Metal Contamination in Silicon Solar Cells*, Adv. Funct. Mater. **21**, 879–890 (2011)

[2] B. O. Kolbesen, W. Bergholz, and H. Wendt, *Impact of Defects on the Technology of Highly Integrated Circuits*, Mater. Sci. Forum **38-41**, 1–12 (1991)

[3] K. Graff, *Metal impurities in silicon-device fabrication*, 2nd, rev. edition, Berlin: Springer (2000)

[4] B. Bathey and M. Cretella, *Solar-grade silicon*, J. Mater. Sci. **17**, 3077–3096 (1982)

[5] A. Goetzberger, C. Hebling, and H. Schock, *Photovoltaic materials, history, status and outlook*, Mat. Sci. Eng. R **40**, 1–46 (2003)

[6] D. Sarti and R. Einhaus, *Silicon feedstock for the multi-crystalline photovoltaic industry*, Sol. Energ. Mat. Sol. C. **72**, 27–40 (2002)

[7] A. Mueller, M. Ghosh, R. Sonnenschein, and P. Woditsch, *Silicon for photovoltaic applications*, Mater. Sci. Eng. B-Solid **134**, 257–262 (2006)

[8] K. Yasuda, K. Morita, and T. H. Okabe, *Processes for Production of Solar-Grade Silicon Using Hydrogen Reduction and/or Thermal Decomposition*, Energy Technology **2**, 141–154 (2014)

[9] A. A. Istratov, H. Hieslmair, and E. R. Weber, *Iron and its complexes in silicon*, Appl. Phys. A **69**, 13–44 (1999)

[10] A. A. Istratov, H. Hieslmair, and E. R. Weber, *Iron contamination in silicon technology*, Appl. Phys. A **70**, 489–534 (2000)

[11] A. A. Istratov and E. R. Weber, *Electrical properties and recombination activity of copper, nickel and cobalt in silicon*, Appl. Phys. A **66**, 123–136 (1998)

[12] H. Kitagawa, *Diffusion and electrical properties of 3d transition-metal impurities in silicon*, Solid State Phenom. **71**, 51–72 (2000)

[13] N. Yarykin and J. Weber, *Copper-related deep-level centers in irradiated p-type silicon*, Phys. Rev. B **83**, 125207 (2011)

[14] M. Nakamura, S. Murakami, and H. Udono, *Transformation reactions of copper centers in the space-charge region of a copper-diffused silicon crystal measured by deep-level transient spectroscopy*, J. Appl. Phys. **112**, 063530 (2012)

[15] M. Nakamura, S. Murakami, and H. Udono, *Copper centers in copper-diffused n-type silicon measured by photoluminescence and deep-level transient spectroscopy*, Appl. Phys. Lett. **101**, 042113 (2012)

[16] N. Yarykin and J. Weber, *Identification of copper-copper and copper-hydrogen complexes in silicon*, Semiconductors **47**, 275–278 (2013)

[17] N. Yarykin and J. Weber, *Deep levels of copper-hydrogen complexes in silicon*, Phys. Rev. B **88**, 085205 (2013)

[18] A. G. Aberle, *Overview on SiN surface passivation of crystalline silicon solar cells*, Sol. Energ. Mat. Sol. C. **65**, 239 – 248 (2001)

[19] C. H. Seager, R. A. Anderson, and J. K. G. Panitz, *The diffusion of hydrogen in silicon and mechanisms for "unintentional" hydrogenation during ion beam processing*. J. Mater. Res. **2**, 96–106 (1987)

[20] J. Weber, S. Knack, O. Feklisova, N. Yarykin, and E. Yakimov, *Hydrogen penetration into silicon during wet-chemical etching*, Microelectron. Eng. **66**, 320 – 326 (2003)

[21] Y. Tokuda, I. Katoh, H. Ohshima, and T. Hattori, *Observation of hydrogen in commercial Czochralski-grown silicon wafers*, Semicond. Sci. Technol. **9**, 1733 (1994)

[22] D. B. Fenner, D. K. Biegelsen, and R. D. Bringans, *Silicon surface passivation by hydrogen termination: A comparative study of preparation methods*, J. Appl. Phys. **66**, 419–424 (1989)

[23] G. S. Higashi, Y. J. Chabal, G. W. Trucks, and K. Raghavachari, *Ideal hydrogen termination of the Si (111) surface*, Appl. Phys. Lett. **56**, 656–658 (1990)

[24] G. W. Trucks, K. Raghavachari, G. S. Higashi, and Y. J. Chabal, *Mechanism of HF etching of silicon surfaces: A theoretical understanding of hydrogen passivation*, Phys. Rev. Lett. **65**, 504–507 (1990)

[25] K. Bergman, M. Stavola, S. J. Pearton, and J. Lopata, *Donor-hydrogen complexes in passivated silicon*, Phys. Rev. B **37**, 2770–2773 (1988)

[26] R. Jones, A. Resende, S. Oberg, and P. Briddon, *The electronic properties of transition metal hydrogen complexes in silicon*, Mater. Sci. Eng. B-Solid **58**, 113–117 (1999)

[27] D. J. Backlund and S. K. Estreicher, *Structural, electrical, and vibrational properties of Ti-H and Ni-H complexes in Si*, Phys. Rev. B **82**, 155208 (2010)

[28] P. Santos, J. Coutinho, V. J. B. Torres, M. J. Rayson, and P. R. Briddon, *Hydrogen passivation of titanium impurities in silicon: Effect of doping conditions*, Appl. Phys. Lett. **105**, 032108 (2014)

[29] W. Jost, J. Weber, and H. Lemke, *Hydrogen-induced defects in cobalt-doped n-type silicon*, Semicond. Sci. Technol. **11**, 22 (1996)

[30] W. Jost, J. Weber, and H. Lemke, *Hydrogen - cobalt complexes in p-type silicon*, Semicond. Sci. Technol. **11**, 525 (1996)

[31] W. Jost and J. Weber, *Titanium-hydrogen defects in silicon*, Phys. Rev B **54**, 11038–11041 (1996)

[32] J.-U. Sachse, *Reaktionen von Wasserstoff mit Übergangsmetallen in Silizium*, Ph.D. thesis, Universität Stuttgart (1997)

[33] R. Jones, S. Öberg, J. Goss, P. R. Briddon, and A. Resende, *Theory of Nickel and Nickel-Hydrogen Complexes in Silicon*, Phys. Rev. Lett. **75**, 2734–2737 (1995)

[34] R. Singh, S. J. Fonash, and A. Rohatgi, *Interaction of low-energy implanted atomic H with slow and fast diffusing metallic impurities in Si*, Appl. Phys. Lett. **49**, 800–802 (1986)

[35] M. Yoneta, Y. Kamiura, and F. Hashimoto, *Chemical etching-induced defects in phosphorus-doped silicon*, J. Appl. Phys. **70**, 1295–1308 (1991)

[36] O. Andersen, A. R. Peaker, L. Dobaczewski, K. Bonde Nielsen, B. Hourahine, R. Jones, P. R. Briddon, and S. Öberg, *Electrical activity of carbon-hydrogen centers in Si*, Phys. Rev. B **66**, 235205 (2002)

[37] A. Endrös, W. Krühler, and J. Grabmaier, *A hydrogen-carbon related deep donor in crystalline n-si-c*, Mater. Sci. Eng. B-Solid **4**, 35–39 (1989)

[38] Y. Kamiura, M. Tsutsue, Y. Yamashita, F. Hashimoto, and K. Okuno, *Deep center related to hydrogen and carbon in p-type silicon*, J. Appl. Phys. **78**, 4478–4486 (1995)

[39] N. M. Johnson, F. A. Ponce, R. A. Street, and R. J. Nemanich, *Defects in single-crystal silicon induced by hydrogenation*, Phys. Rev. B **35**, 4166–4169 (1987)

[40] Y. L. Huang, E. Simoen, C. Claeys, J. M. Rafi, and P. Clauws, *A DLTS study on plasma-hydrogenated n-type high-resistivity magnetic Cz silicon*, J. Mater. Sci. - Mater. Electron. **18**, 705–710 (2007)

[41] O. Vyvenko, N. Bazlov, M. Trushin, A. Nadolinski, M. Seibt, W. Schroter, and G. Hahn, *Impact of hydrogenation on electrical properties of $NiSi_2$ precipitates in silicon*, Solid State Phenom. **108-109**, 279–284 (2005)

[42] L. Tsetseris, S. Wang, and S. T. Pantelides, *Thermal donor formation processes in silicon and the catalytic role of hydrogen*, Appl. Phys. Lett. **88**, 051916 (2006)

[43] E. Simoen, C. Claeys, J. Rafí, and A. Ulyashin, *Thermal donor formation in direct-plasma hydrogenated n-type Czochralski silicon*, Mater. Sci. Eng. B-Solid **134**, 189 – 192 (2006)

[44] K. Bonde Nielsen, L. Dobaczewski, S. Søgård, and B. Bech Nielsen, *Electronic levels of isolated and oxygen-perturbed hydrogen in silicon and migration of hydrogen*, Physica B **308-310**, 134 – 138 (2001)

[45] K. B. Nielsen, L. Dobaczewski, S. Søgård, and B. B. Nielsen, *Acceptor state of monoatomic hydrogen in silicon and the role of oxygen*, Phys. Rev. B **65**, 075205 (2002)

[46] C. Herring, N. M. Johnson, and C. G. Van de Walle, *Energy levels of isolated interstitial hydrogen in silicon*, Phys. Rev. B **64**, 125209 (2001)

[47] S. Muto, S. Takeda, and M. Hirata, *Hydrogen-induced platelets in silicon studied by transmission electron microscopy*, Philos. Mag. A **72**, 1057–1074 (1995)

[48] J. N. Heyman, J. W. Ager, E. Haller, N. M. Johnson, J. Walker, and C. M. Doland, *Hydrogen-induced Platelets In Silicon - Infrared-absorption And Raman-scattering*, Phys. Rev. B **45**, 13363–13366 (1992)

[49] E. V. Lavrov and J. Weber, *Evolution of Hydrogen Platelets in Silicon Determined by Polarized Raman Spectroscopy*, Phys. Rev. Lett. **87**, 185502 (2001)

[50] N. Fukata, S. Sasaki, K. Murakami, K. Ishioka, M. Kitajima, S. Fujimura, and J. Kikuchi, *Formation of Hydrogen Molecules in n-Type Silicon*, Jpn. J. Appl. Phys. **35**, L1069–L1071 (1996)

[51] R. E. Pritchard, M. J. Ashwin, J. H. Tucker, and R. C. Newman, *Isolated interstitial hydrogen molecules in hydrogenated crystalline silicon*, Phys. Rev. B **57**, R15048–R15051 (1998)

[52] P. Blood and J. Orton, *The Electrical Characterization of Semiconductors: Majority Carriers and Electron States*, London: Academic Press (1992)

[53] O. Feklisova and N. Yarykin, *Transformation of deep-level spectrum of irradiated silicon due to hydrogenation under wet chemical etching*, Semicond. Sci. Technol. **12**, 742–749 (1997)

[54] C. H. Henry and D. V. Lang, *Nonradiative capture and recombination by multiphonon emission in gaas and gap*, Phys. Rev. B **15**, 989–1016 (1977)

[55] H. Goto, Y. Adachi, and T. Ikoma, *Cross-section of multiphonon-emission carrier capture at deep centers in compound semiconductors*, Phys. Rev. B **22**, 782–796 (1980)

[56] K. Peuker, R. Enderlein, A. Schenk, and E. Gutsche, *Theory of nonradiative multiphonon capture processes - solution of old controversies*, Phys. Status Solidi B **109**, 599–606 (1982)

[57] J. Bourgoin and M. Lannoo, *Point Defects in Semiconductors II*, Berlin: Springer (1983)

[58] J. Hartke, *3-dimensional poole-frenkel effect*, J. Appl. Phys. **39**, 4871 (1968)

[59] M. Levinshtein, S. Rumyantsev, and M. Shur, *The Handbook Series on Semiconductor Parameters I*, World Scientific (1996)

[60] J. Frenkel, *On Pre-Breakdown Phenomena in Insulators and Electronic Semi-Conductors*, Phys. Rev. **54**, 647–648 (1938)

[61] S. D. Ganichev, E. Ziemann, W. Prettl, I. N. Yassievich, A. A. Istratov, and E. R. Weber, *Distinction between the Poole-Frenkel and tunneling models of electric-field-stimulated carrier emission from deep levels in semiconductors*, Phys. Rev. B **61**, 10361–10365 (2000)

[62] V. Karpus and V. I. Perel', *Multiphoton ionization of deep centers in semiconductors in an electric field*, Zh. Eksp. Teor. Fiz. **91**, 2319 (1986) [Sov. Phys. JETP **64**, 1376 (1986)]

[63] D. V. Lang, *Deep-level transient spectroscopy: A new method to characterize traps in semiconductors*, J. Appl. Phys. **45**, 3023–3032 (1974)

[64] A. Wang and C. Sah, *Determination of trapped charge emission rates from nonexponential capacitance transients due to high trap densities in semiconductors*, J. Appl. Phys. **55**, 565–570 (1984)

[65] L. Dobaczewski, A. R. Peaker, and K. B. Nielsen, *Laplace-transform deep-level spectroscopy: The technique and its applications to the study of point defects in semiconductors*, J. Appl. Phys. **96**, 4689–4728 (2004)

[66] H. Lefevre and M. Schulz, *Double correlation technique (DDLTS) for analysis of deep level profiles in semiconductors*, Appl. Phys. **12**, 45–53 (1977)

[67] W. Schroter, J. Kronewitz, U. Gnauert, F. Riedel, and M. Seibt, *Band-like and localized states at extended defects in silicon*, Phys. Rev. B **52**, 13726–13729 (1995)

[68] W. von Ammon, R. Hölzl, J. Virbulis, E. Dornberger, R. Schmolke, and D. Gräf, *The impact of nitrogen on the defect aggregation in silicon*, J. Cryst. Growth **226**, 19 – 30 (2001)

[69] A. Karoui, F. Karoui, G. Rozgonyi, M. Hourai, and K. Sueoka, *Structure, energetics, and thermal stability of nitrogen-vacancy-related defects in nitrogen doped silicon*, J. Electrochem. Soc. **150**, G771–G777 (2003)

[70] V. Voronkov and R. Falster, *Nitrogen interaction with vacancies in silicon*, Mater. Sci. Eng. B-Solid **114-115**, 130 – 134 (2004)

[71] P. Saring, private communication

[72] S. Pearton, J. Kahn, and E. Haller, *Deep level effects in silicon and germanium after plasma hydrogenation*, J. Electron. Mater. **12**, 1003–1014 (1983)

[73] R. Stübner, V. Kolkovsky, and J. Weber, *Identification of the metastable carbon-hydrogen complex in silicon*, J. Appl. Phys. , *submitted* (2014)

[74] R. Newman, *Infra-red absorption due to localized modes of vibration of impurity complexes in ionic and semiconductor crystals*, Adv. Phys. **18**, 545–663 (1969)

[75] T. M. Gibbons, S. K. Estreicher, K. Potter, F. Bekisli, and M. Stavola, *Huge isotope effect on the vibrational lifetimes of an $H_2^*(C)$ defect in Si*, Phys. Rev. B **87**, 115207 (2013)

[76] D. West and S. Estreicher, *Isotope dependence of the vibrational lifetimes of light impurities in Si from first principles*, Phys. Rev. B **75**, 075206 (2007)

[77] K. Kohli, G. Davies, N. Vinh, D. West, S. Estreicher, T. Gregorkiewicz, I. Izeddin, and K. Itoh, *Isotope Dependence of the Lifetime of the* $1136\text{-}cm^{-1}$ *Vibration of Oxygen in Silicon*, Phys. Rev. Lett. **96**, 225503 (2006)

[78] S. Hocine and D. Mathiot, *Titanium diffusion in silicon*, Appl. Phys. Lett. **53**, 1269–1271 (1988)

[79] J. T. Borenstein, J. I. Hanoka, B. R. Bathey, J. P. Kalejs, and S. Mil'shtein, *Influence of ion-implanted titanium on the performance of edge-defined, film-fed grown silicon solar cells*, Appl. Phys. Lett. **62**, 1615–1616 (1993)

[80] J.-W. Chen, A. Milnes, and A. Rohatgi, *Titanium in silicon as a deep level impurity*, Solid-State Electronics **22**, 801–808 (1979)

[81] J. Morante, J. Carceller, P. Cartujo, and J. Barbolla, *Thermal emission rates and capture cross-section of majority carriers at titanium levels in silicon*, Solid-State Electronics **26**, 1–6 (1983)

[82] A. C. Wang and C. T. Sah, *Complete electrical characterization of recombination properties of titanium in silicon*, J. Appl. Phys. **56**, 1021–1031 (1984)

[83] D. Mathiot and S. Hocine, *Titanium-related deep levels in silicon: A reexamination*, J. Appl. Phys. **66**, 5862–5867 (1989)

[84] S. Leonard, V. P. Markevich, A. R. Peaker, and B. Hamilton, *Passivation of titanium by hydrogen in silicon*, Appl. Phys. Lett. **103**, 132103 (2013)

[85] K. Bonde Nielsen, B. B. Nielsen, J. Hansen, E. Andersen, and J. U. Andersen, *Bond-centered hydrogen in silicon studied by in situ deep-level transient spectroscopy*, Phys. Rev. B **60**, 1716–1728 (1999)

[86] V. P. Markevich, S. Leonard, A. R. Peaker, B. Hamilton, A. G. Marinopoulos, and J. Coutinho, *Titanium in silicon: Lattice positions and electronic properties*, Appl. Phys. Lett. **104**, 152105 (2014)

[87] C. Van de Walle and J. Neugebauer, *Universal alignment of hydrogen levels in semi-conductors, insulators and solutions*, Nature **423**, 626–628 (2003)

[88] H. Lemke, *Energy-level properties of cobalt in silicon*, Phys. Status Solidi A **85**, K133–K136 (1984)

[89] H. Kitagawa, H. Nakashima, and K. Hashimoto, *Energy-levels and solubility of electrically active cobalt in silicon studied by combined hall and dlts measurements*, Jpn. J. Appl. Phys. **24**, 373–374 (1985)

[90] H. Nakashima, Y. Tsumori, T. Miyagawa, and K. Hashimoto, *Deep impurity levels of cobalt in silicon*, Jpn. J. Appl. Phys. **29**, 1395–1398 (1990)

[91] E. Scheibe and W. Schröter, *Investigations in cobalt doped silicon by DLTS and Mössbauer effect*, Physica B **116**, 318 – 322 (1983)

[92] W. Bergholz and W. Schröter, *Precipitation of cobalt in silicon studied by Mössbauer spectroscopy*, Phys. Status Solidi A **49**, 489–498 (1978)

[93] Z. Z. Zhang, B. Partoens, K. Chang, and F. M. Peeters, *First-principles study of transition metal impurities in Si*, Phys. Rev. B **77**, 155201 (2008)

[94] F. Beeler, O. K. Andersen, and M. Scheffler, *Electronic and magnetic structure of 3d-transition-metal point defects in silicon calculated from first principles*, Phys. Rev. B **41**, 1603–1624 (1990)

[95] H. Lemke and K. Irmscher, *Proof of Interstitial Cobalt Defects in Silicon Float Zone Crystals Doped during Crystal Growth*, ECS Trans. **3**, 299–310 (2006)

[96] M. Pasemann, W. Bergholz, and W. Schröter, *Mössbauer Spectroscopical and Electron Microscopical Investigations of the Behaviour of Cobalt in Silicon*, Phys. Status Solidi A **81**, 273–280 (1984)

[97] W. Bergholz, *Pairing reactions of interstitial cobalt and shallow acceptors in silicon observed in Mössbauer spectroscopy*, Physica B **116**, 312 – 317 (1983)

[98] D. Mathiot, *Cobalt related levels in P and (P+B) doped n-type silicon - possible observation of the (CoB) pair*, J. Appl. Phys. **65**, 1554–1558 (1989)

[99] A. J. Tavendale, D. Alexiev, and A. A. Williams, *Field drift of the hydrogen-related, acceptor-neutralizing defect in diodes from hydrogenated silicon*, Appl. Phys. Lett. **47**, 316–318 (1985)

[100] Y. Tokumaru, *Properties of Silicon Doped with Nickel,* Jpn. J. Appl. Phys. **2**, 542 (1963)

[101] H. Kitagawa and H. Nakashima, *Nickel-related deep levels in silicon studied by combined hall effect and DLTS measurement,* Phys. Status Solidi A **99**, K49–K52 (1987)

[102] H. Lemke, *Dotierungseigenschaften von Nickel in Silizium,* Phys. Status Solidi A **99**, 205–213 (1987)

[103] H. Kitagawa, S. Tanaka, H. Nakashima, and M. Yoshida, *Electrical properties of nickel in silicon,* J. Electron. Mater. **20**, 441–447 (1991)

[104] M. Shiraishi, J. U. Sachse, H. Lemke, and J. Weber, *DLTS analysis of nickel-hydrogen complex defects in silicon,* Mater. Sci. Eng. B-Solid **58**, 130 – 133 (1999)

[105] S. Tanaka and H. Kitagawa, *Distribution of electrically active nickel atoms in silicon crystals measured by means of deep level transient spectroscopy,* Physica B **401-402**, 115 – 118 (2007)

[106] B. Effey-Schwickert, M. Wiegand, H. Vollmer, and R. Labusch, *Nickel in silicon studied by electron paramagnetic resonance,* Appl. Phys. A **77**, 711–716 (2003)

[107] D. J. Backlund and S. K. Estreicher, *Ti, Fe, and Ni in Si and their interactions with the vacancy and the A center: A theoretical study,* Phys. Rev. B **81**, 235213 (2010)

[108] H. Indusekhar and V. Kumar, *Electrical properties of nickel-related deep levels in silicon,* J. Appl. Phys. **61**, 1449–1455 (1987)

[109] K. Murakami, H. Kuribayashi, and K. Masuda, *Electronic-energy level of off-center substitutional nitrogen in silicon - determination by electron-spin resonance measurements,* Jpn. J. Appl. Phys. **27**, L1414–L1416 (1988)

[110] G. Voronkova, A. Batunina, L. Moiraghi, V. Voronkov, R. Falster, and M. Milvidski, *Deep level generation in nitrogen-doped float-zoned silicon,* Nucl. Instrum. Meth. B **253**, 217 – 221 (2006)

[111] W. R. Buchwald and N. M. Johnson, *Revised role for the Poole–Frenkel effect in deep-level characterization,* J. Appl. Phys. **64**, 958–961 (1988)

[112] V. Kveder, W. Schroter, M. Seibt, and A. Sattler, *Electrical activity of dislocations in Si decorated by Ni,* Solid State Phenom. **82-84**, 361–366 (2002)

[113] M. Seibt and V. Kveder, *Microstructural and electrical properties of NiSi$_2$ precipitates at dislocations in silicon*, Solid State Phenom. **95-96**, 447–452 (2004)

[114] A. Peaker, J. Evans-Freeman, P. Kan, L. Rubaldo, I. Hawkins, K. Vernon-Parry, and L. Dobaczewski, *Hydrogen reactions with electron irradiation damage in silicon*, Physica B **273-274**, 243 – 246 (1999)

[115] K. Bonde Nielsen, L. Dobaczewski, K. Goscinski, R. Bendesen, O. Andersen, and B. Bech Nielsen, *Deep levels of vacancy-hydrogen centers in silicon studied by Laplace DLTS*, Physica B **273-274**, 167 – 170 (1999)

[116] K. A. Kikoin and V. N. Fleurov, *Transition Metal Impurities in Semiconductors*, World Scientific (1994)

[117] P. Kaminski, R. Kozlowski, A. Jelenski, T. Mchedlidze, and M. Suezawa, *High-Resolution Photoinduced Transient Spectroscopy of Electrically Active Iron-Related Defects in Electron Irradiated High-Resistivity Silicon*, Jpn. J. Appl. Phys. **42**, 5415 (2003)

[118] H. Lemke, *Störstellenreaktionen bei Cu-dotierten Siliziumkristallen*, Phys. Status Solidi A **95**, 665–677 (1986)

[119] G. Weyer, A. Burchard, M. Fanciulli, V. Fedoseyev, H. Gunnlaugsson, V. Mishin, and R. Sielemann, *The electronic configuration of substitutional Fe in silicon*, Physica B **273-274**, 363 – 366 (1999)

[120] D. J. Silva, U. Wahl, J. G. Correia, and J. P. Araújo, *Influence of n+ and p+ doping on the lattice sites of implanted Fe in Si*, J. Appl. Phys. **114**, 103503 (2013)

[121] A. Zunger and U. Lindefelt, *Theory of substitutional and interstitial 3d impurities in silicon*, Phys. Rev. B **26**, 5989–5992 (1982)

[122] A. Istratov and E. Weber, *Physics of copper in silicon*, J. Electrochem. Soc. **149**, G21–G30 (2002)

[123] S. Knack, J. Weber, and H. Lemke, *Hydrogen-rhodium complexes in silicon*, Mater. Sci. Eng. B-Solid **58**, 141 – 145 (1999)

[124] J. Bollmann, S. Knack, J. Weber, and ISOLDE collaboration, *Iridium-Related Deep Levels in n-Type Silicon*, Phys. Status Solidi B **222**, 251–260 (2000)

[125] J.-U. Sachse, J. Weber, and H. Lemke, *Deep-level transient spectroscopy of Pd-H complexes in silicon*, Phys. Rev. B **61**, 1924–1934 (2000)

[126] J.-U. Sachse, E. Ö. Sveinbjörnsson, W. Jost, J. Weber, and H. Lemke, *Electrical properties of platinum-hydrogen complexes in silicon*, Phys. Rev. B **55**, 16176–16185 (1997)

[127] J. Chen, Vl. Kolkovsky, and J. Weber, *New Pt-H complexes in silicon*, Unpublished results

[128] J.-U. Sachse, E. Ö. Sveinbjörnsson, N. Yarykin, and J. Weber, *Similarities in the electrical properties of transition metal-hydrogen complexes in silicon*, Mater. Sci. Eng. B-Solid **58**, 134–140 (1999)

[129] F. Beeler and M. Scheffler, *Theory of 4d Transition-Metal Ions in Silicon: Total-Energies, Diffusion, Electronic and Magnetic Properties*, Mater. Sci. Forum **38-41**, 257 (1991)

Danksagungen

Viele Personen haben mich während meiner Promotion unterstützt. Besonders danken möchte ich

Prof. Dr. Jörg Weber für die Möglichkeit, an seinem Institut zu promovieren, und für den unermüdlichen Ansporn, physikalischen Ungereimtheiten auf den Grund zu gehen.

Dr. Vladimir Kolkovsky für die herausragende Betreuung, für die er auch seine Freizeit geopfert hat.

Dir. of Res. Dr. Abdelmadjid Mesli for the stimulating discussions during his stay in Dresden and for writing the second report.

Philipp Saring für die freundliche Bereitstellung der eindiffundierten Nickel-Proben.

Dr. Edward Lavrov für die guten Gespräche und die vielen Rätsel- und Knobelaufgaben zur Geistesauflockerung.

Dr. Ellen Hieckmann für REM-Bilder und die gute Atmosphäre im Institut.

Dr. Teimuraz Mtchedlidze for the good atmosphere in the lab and the scientific discussions.

meinen Mitdoktoranden Sandro, Matthias, Ronald, Dirk, Sebastian, Frank und Karsten für die wissenschaftlichen und sonstigen Diskussionen und die gute Atmosphäre im Institut.

Berthold Köhler für Hilfe und Anleitung bei technischen und handwerklichen Problemen aller Art.

Simone Behrendt für das bereitwillige Zusammenstellen unheimlicher chemischer Mixturen.

Steffi Gerber für ihre Hilfe bei allen bürokratischen Anliegen und für die großartige Organisation von Wandertagen und Feiern.

meiner Familie für ihre Unterstützung und dass sie für mich da ist.